Assessing Technology Gaps for the Federal Highway Administration Digital Highway Measurement Program

FHWA Grant DTFH61-09-X-30011
NISTIR 7685

Final Report
April 2010

Submitted to:

Federal Highway Administration
Office of Safety Research and Development
Turner-Fairbank Highway Research Center
6300 Georgetown Pike
McLean, VA 22101

Submitted by:

Geraldine Cheok, Marek Franaszek, Itai Katz, Alan Lytle, Kamel Saidi, Nicholas Scott
Construction Metrology and Automation Group, Building and Fire Research Laboratory
National Institute of Standards and Technology
100 Bureau Drive
Gaithersburg, MD 20899

Disclaimer

Certain commercial equipment, instruments, or materials are identified in this document. Such identification does not imply recommendation or endorsement by the National Institute of Standards and Technology, nor does it imply that the products identified are necessarily the best available for the purpose.

Table of Contents

Disclaimer i
Table of Contents iii
List of Figures v
List of Tables vii
List of Acronyms and Abbreviations ix
Executive Summary xi
1. Introduction 1
    1.1 Background 1
    1.2. Project Objectives and Tasks 4
    1.3. Report Structure 5
2. Master Element Table 7
3. Technology Table 10
4. Technology Gap Assessment 22
    4.1 Definitions 23
        4.1.1 Demand 23
        4.1.2 Maturity Index 24
        4.1.3. Impact 27
        4.1.4. Technology Readiness Level Subcategories 29
    4.2. Technology Functional Classification Categories 29
        4.2.1. Location Measurement 29
        4.2.2. Object and Feature Recognition 30
        4.2.3 Sign Type Recognition 31
        4.2.4. Character Recognition 31
        4.2.5. Spatial Measurement 31
        4.2.6. Material Optical Property Measurement 32
        4.2.7. Material Property Measurement 32
    4.3 Determining High, Medium and Low Priority Elements 33
    4.4. Demand Calculation 34
    4.5 Maturity Index Calculation 36
    4.6. Results 38
5. Recommendations 41
    5.1 Recommendation One 41
    5.2 Recommendation Two 42
    5.3 Recommendation Three 45
Acknowledgements 47
References 47

Appendix A: Digital Highway Measurement Technology Database 49
Appendix B: Technology Search Procedure 56
Appendix C: Element Priority Survey 59
Appendix D: FHWA List of Element Priorities 79
    D.1 Background 79
    D.2 Criteria for Defining High Priority Data Elements 80
    D.3 Criteria for Defining Medium Priority Data Elements 82
    D.4 Criteria for Defining Low Priority Data Elements 84
Appendix E: Combined List of Element Priorities 85
Bibliography 94

List of Figures

Figure 3.1. Percentage of Addressed elements within each MET category. 14

Figure 3.2. Number of technologies in each of the technology functional classification categories. 15

Figure 3.3. Number of identified technologies grouped by general technology category. 16

Figure 3.4. Number of elements addressed (black) and potentially addressed (white) by technologies T001 to T072. 17

Figure 3.5. Number of elements addressed (black) and potentially addressed (white) by technologies T073 to T143. 18

Figure 3.6. Number of technologies within each functional classification category that provide real-time, offline, and unknown data processing speed. 19

Figure 3.7. Number of technologies within each functional classification category that provide automated, semi-automated, manual, and unknown levels of data processing automation. 20

Figure 4.1. Schematic of the Demand vs. Maturity Graph. 22

Figure 4.2. Framework for the analysis procedure. 23

Figure 4.3. Impact as defined by the Demand vs. Maturity Plot. 28

Figure 4.4. Number of elements belonging to each technology functional classification category. 36

Figure 4.5. Demand vs. Maturity Graph. 39

Figure 5.1. The Receiver Operating Curve (ROC) characterizes the performance of a classifier. 44

List of Tables

Table 1.	Data Collected by DHMS and State-of-practice of Technology	3
Table 2.	Description of the Data Fields in the MET	9
Table 3.	Description of the Data Fields in the Technology Table	12
Table 4.	Number of High, Medium and Low Priority Elements and Demands	35
Table 5.	Maturity Index Calculation	37
Table 6.	Sensitivity of Maturity Index to Additional COTS Technologies Found	38
Table C.1.	High Priority Elements in Survey	59
Table C.2.	Survey Results	67
Table D.1.	List of FHWA High Priority Elements	81
Table D.2.	List of FHWA Medium Priority Elements	82

List of Acronyms and Abbreviations

COTS	Commercial-off-the-Shelf
DHMS	Digital Highway Measurement System
DMI	Distance Measuring Instrument
FHWA	Federal Highway Administration
GPS	Global Positioning System
HPMS	Highway Performance Monitoring System
IHSDM	Interactive Highway Safety Design Model
INS	Inertial Navigation System
LADAR	Laser Detection and Ranging
LIDAR	Light Detection and Ranging
MET	Master Element Table
MIRE	Minimum Inventory of Roadway Elements
MMIRE	Model Minimum Inventory of Roadway Elements
MMUCC	Model Minimum Uniform Crash Criteria
MUTCD	Manual on Uniform Traffic Control Devices
NCDOT	North Carolina Department of Transportation
SHRP-2	Strategic Highway Research Program
SIFT	Scale Invariant Feature Transform
TSIMS	Transportation Safety Information Management System
TT	Technology Table
US&R	Urban Search and Rescue
WAAS	Wide Area Augmentation System

Executive Summary

The ability to automatically collect more and better measurements of roadway assets (e.g., existence, condition, location, type) and to accurately map both geometry and elements of roadway assets is an important mitigating factor in roadway safety. This information serves numerous purposes including verification of road design codes, detecting degradation in roadway elements, detecting roadside hardware and clutter changes which impact safety (e.g., sight lines), creation of a roadway element inventory, analysis of accident data correlated to roadway inventory, and simulation of roadway conditions for safety analysis. To address the need to collect roadway information economically, accurately, and reliably, the Federal Highway Administration's (FHWA) Office of Advanced Research initiated the development of the Digital Highway Measurement System (DHMS) in 2003. The DHMS is a vehicle equipped with multiple sensors and instrumentation to collect road environment (roadway and roadside) geometry, roadside inventory, and pavement condition data at highway speeds. The DHMS vision is to enable, using state-of-the-art sensors and analysis, the collection of roadway geometry at required levels of accuracy not achievable by the current state-of-practice, and to introduce methods and metrics for capturing the presence, location, and health of all roadside assets.

To support future DHMS technology development, FHWA requires identification of technology gaps between currently available data collection technologies and the need for capabilities to efficiently collect comprehensive information about the nation's roadway infrastructure. FHWA also requires a prioritization of technology gaps to future investment in DHMS research and development. The objectives of this project are to:

Create a master element table of roadway elements.

The Master Element Table (MET) is a consolidation of the roadway elements available in the Model Minimum Inventory of Roadway Elements (MMIRE) (Council, Harkey, Carter, & White, 2007), the element lists generated for the SHRP-2 (Strategic Highway Research Program) S-03 rodeo and the North Carolina Asset Management Workshop, the National Bridge Index, and the U.S. National Highway-Rail Crossing Inventory Program (Federal Railroad Administration, 2007). The MET is limited to above-ground elements. Each element was reviewed to understand the actual physical characteristic data that required measurement in order to extract the relevant information needed.

Conduct a literature review/market analysis study to identify relevant technologies.

A literature review/market analysis survey was conducted for each of the MET elements to a) identify relevant technologies, b) perform an assessment of the readiness of that technology (where information was available), and c) document providers and researchers involved with the technology delivery. A Technology Table was created from the search for technologies in which applicable technologies were listed. These technologies were tagged with the following meta-data:

- What elements would this technology address?
- What is the estimated technology readiness? (e.g., commercial-off-the-shelf (COTS), research & development (R&D), Prototype/ Demonstration, Concept).
- What is the data processing speed? (e.g., real-time, offline post-processing).
- What is the data processing automation level? (e.g., automated, semi-automated, manual).

- What general category of technology does the specific technology utilize? (e.g., the technology 'uses -' laser-based 3D imaging, camera-based 3D imaging, GPS, inertial systems, software only (mainly for research algorithms), laser-based 2D imaging, camera-based 2D imaging, or other).
- What general functional capability does the specific technology provide? (e.g., the technology 'provides -' location measurement, object or feature recognition, sign-type recognition, character recognition, spatial measurement, material property optical measurement, or material property measurement).

Develop an objective process to identify and prioritize technology gaps.

Potential technology gaps were identified by analyzing the demand and maturity of a technology grouping. The technologies were grouped by functional classification (i.e., what does the technology provide?): location measurement, object/feature recognition, sign-type recognition, character recognition, spatial measurement, material optical property measurement, and material property measurement. Demand is defined as the weighted sum of the high, medium and low priority elements within a technology functional classification category divided by the weighted sum of the total number of high, medium, and low priority elements in-scope. The Maturity Index of a technology functional classification category is defined as the weighted sum of the maturity indices for the category's technology readiness level, level of automation, data processing speed, and the ability to measure an element to the required accuracy or resolution. Impact is defined by a region in the Demand vs. Maturity plot. In general, the relative locations of technology functional classification categories in the Demand vs. Maturity plot indicates the impact of the category relative to the other categories. The category with the highest demand and lowest maturity would have the highest impact and the category with the lowest demand and highest maturity would have the lowest impact.

Using the approach described in the previous paragraph, technologies within the functional category of Object and Feature Recognition were rated as the highest priority and the functional category of Spatial Measurement as the second highest priority. The three recommendations were:

- Develop performance-based standards for digital highway measurement
- Provide standard reference data sets for DHMS algorithm development
- Develop object/feature recognition algorithms

Although not specifically part of the technology gap assessment, development of open, performance-based standards and the creation of standard reference data sets are highly recommended and would likely provide the highest return on investment for FHWA in advancing DHMS technologies.

1. Introduction

1.1 Background

According to data compiled by the National Highway Traffic Safety Administration (NHTSA), there were 6,024,000 police-reported motor vehicle traffic crashes in 2007. Of those crashes, 29 % resulted in 2,491,000 persons injured and 41,059 deaths. The remaining 4,275,000 accidents involved only property damage (NHTSA, 2009). Although there are no reported economic figures associated with these data, a comparison can be made to the year 2000, during which there were 6,394,000 police-reported motor vehicle crashes resulting in 2,070,000 injuries and 37,409 fatalities (NHTSA, 2000). The economic impact of motor vehicle crashes in 2000 was $230.6 billion (Blincoe, et al., 2002). Accidents remain a leading cause of death in the U.S. (#5 in 2006), and motor vehicle accidents account for nearly 25 % of those fatalities (Heron, et al., 2009).

Approximately one-third of motor vehicle traffic crashes have road geometry and roadway infrastructure as a contributing factor (Rumar, 1985). The ability to inventory and to measure the existing conditions of roadway assets and to accurately map those elements in relation to the geometry is important to improving roadway safety. This information serves numerous purposes including verifying if road design is within specifications, detecting degradation in roadway elements, detecting roadside hardware and clutter changes which impact safety (e.g., sight lines), and creation of a roadway element inventory, among others.

To address the need to collect roadway information economically, accurately, and reliably, FHWA's Office of Advanced Research initiated the development of the Digital Highway Measurement System (DHMS) in 2003. The DHMS is a vehicle equipped with multiple sensors and instrumentation to collect road environment (roadway and roadside) geometry, roadside inventory, and pavement condition data at highway speeds. The DHMS vision is to enable, using state-of-the-art sensors and analysis, the collection of roadway geometry at required levels of accuracy not achievable by the current state-of-practice, and to introduce methods and metrics for capturing the presence, location, and health of all roadside assets. The primary drivers for the DHMS are (Trentacoste, 2006):

- Provide necessary roadway data to conduct safety analyses
- Meet asset management requirements for road safety and condition assessment (GASB-34[1])
- Enable the integration of safety data with infrastructure data
- Enable cost-effective and safe data collection through automation
- Examine next generation sensors (e.g., ground penetrating radar) for the collection of enhanced road data

The central purpose of the DHMS is to support the commitment of highway agencies to make data-driven decisions about safety improvements. Through legislative and conventional educational means, state and local agencies are being encouraged to collect more data, to improve the quality of their data, and to better use their data in order to make the best decisions on how to use available funds to make road systems safer and more efficient. The DHMS is used to identify and demonstrate technical means of collecting and processing these data. Demonstrating these capabilities is a key method of technology transfer and also supports the future development of strong data collection performance standards. Technology transfer is further enabled through cooperative research and development and licensing of DHMS technologies through standard Federal Acquisition Regulation (FAR) provisions (Cobb, 2009).

---

[1] GASB-34 is the Governmental Accounting Standards Board Statement No. 34, which mandated that state and local governments report on the value of infrastructure assets (e.g., roads, bridges, etc.), and develop methods for managing those assets. (DOT, 2000)

Table 1, from FHWA, lists the information that the ideal data collection system would gather. The list is presented generally in increasing order of value, and increasing order of difficulty or complexity. The table also indicates FHWA's estimate of the current state-of-practice within the industry. FHWA's ultimate vision of a data collection system would assess both "How ..." and "When to repair or upgrade ...," with relevant information extracted and delivered in real-time[2] as roadway measurement took place (Cobb, 2009).

Table 1. Data Collected by DHMS and State-of-practice of Technology

	Item	State of Industry/Practice (0=Not available; 4=Mature, readily available)	Comments
a	Where it is.	4	
b	What it is.	3	Optical means effective, but automated object extraction not robust.
c	Its distinguishing quantities, measurements and/or characteristic values. (How much, how many, what type, etc.)	2	Not all identifiable features can be measured directly. Some measurements are not accurate or precise.
d	How well it's working, or meets standards.	1	Only subjective assessment of video.
e	Whether it needs repairs/upgrades	1	Only subjective assessment of video.
f	How to repair or upgrade it.	0	Holy Grail!
g	When to repair or upgrade it.	0	

- "How well it's working, or meets standards." – Assesses remaining functionality of hardware and etc., (e.g., how much of its nominal performance does a damaged guardrail retain?). Alternately, determine whether current condition meets

---

[2] Real-time is a subjective qualifier. The important aspect is that information is delivered to end-users as quickly as possible after roadway measurement. The less delay for post-processing the better with an ultimate vision that information is extracted and delivered wirelessly while the measurement platform is in the field.

established standards (e.g. whether current retro-reflectivity of a sign meets minimum requirements).

- "Whether it needs repairs/upgrades." – Indicates whether the extent of damage reaches the threshold for intervention.

- "How to repair or upgrade it?" – Indicates to users what methods of intervention are needed, including materials needed, repair protocols to be used, etc. For example, a moderately bent guardrail may need only one replacement w-beam rail, one block-out, and one set of splice fasteners.

- "When to repair or upgrade it?" – Indicates the ideal time to make a repair based on lowest overall maintenance cost, or on minimizing construction time, etc. For example, some pavement surface distress may be a clear harbinger of worse to come, but may not represent an immediate hazard. On the other hand, there will come a time when deterioration has gone so far that repair costs increase dramatically. The ideal system will be capable of providing a schedule of needed repairs looking forward a year or more.

The project described in this report is designed to assist the FHWA Office of Safety Research and Development in crafting a research roadmap for future DHMS investment.

1.2. Project Objectives and Tasks

FHWA requires identification of technology gaps between currently available data collection technologies and the need for capabilities to efficiently collect comprehensive information about the nation's roadway infrastructure. Associated with this need is a prioritization of research gaps to guide FHWA's future investment in Digital Highway Measurement System (DHMS) research and development.

The Office of Safety Research commissioned this project to conduct a review of data collection technologies to provide an understanding and mapping of sensing technologies (hardware and software) to the list of roadway elements available in the Model Minimum Inventory of Roadway Elements (MMIRE) (Council, Harkey, Carter, & White, 2007), SHRP-2 S-03 rodeo (Vandervalk, 2008), and the North Carolina Asset Management workshop. The following tasks outline the steps used to accomplish the project objectives:

- Create a master element table of the roadway elements available in the MMIRE, and the element lists generated for the SHRP-2 S-03 rodeo and the North Carolina Asset Management Workshop. The table will be limited to above-ground elements. Each retained element will be reviewed to understand the actual physical characteristic data that need to be measured in order to extract the relevant information needed.

- Conduct a literature review/market analysis study to identify relevant technologies, assess the readiness of those technologies, document the providers and researchers involved with the technology delivery, and identify potential technology gaps.

- Develop and implement an objective process for prioritizing technology gaps identified during the literature review/market analysis study. Provide and discuss a prioritized list of technology gaps.

1.3. Report Structure

Section 2 provides a description of the generation of the Master Element Table (MET), including element selection criteria. Section 3 presents the results of the literature review/market analysis survey conducted on each of the elements from the MET and the resulting Technology Table (TT). Section 4 details the methods used to identify the broad category technology gaps through analysis of the MET and TT. Recommendations are provided in Section 5. Appendix A provides a description of the database that contains the MET and TT and provides examples of the reports that can be generated from the database. Appendix B lists the procedures to conduct the technology search and Appendix C presents the survey questionnaire that was sent to subject matter experts at FHWA as well as the results of the survey. Appendix D includes prioritized lists of high-priority and medium priority elements provided by FHWA and the criteria that were used to generate them. Finally, Appendix E provides the combined lists of high, medium, and low priority elements from the survey (Appendix C) and the FHWA lists (Appendix D).

2. Master Element Table

The Master Element Table (MET) was generated by combining data elements from the MMIRE, SHRP-2 S-03 Rodeo, the North Carolina DOT (NCDOT) Inventory Asset Data Collection (NCDOT), and the NCDOT Pavement Condition Survey (NCDOT). There were 180 data elements from the MMIRE, 120 from SHRP-2, and 134 from NCDOT, 23 data elements from the National Bridge Index (FHWA, 1995), and 38 data elements from the U.S. National Highway-Rail Crossing Inventory Program (Federal Railroad Administration, 2007) bringing the total number of elements in the MET to 495. All of the elements were reviewed to remove duplicate or similar data elements. This process reduced the number of data elements from 495 to 392. A further sub-selection of the 392 elements was made to determine the technology gaps (see Section 4).

The categorization of the data elements in the MMIRE document differed from that in the SHRP-2 and NCDOT documents. The decision was made to follow the MMIRE categorization and to add three new subcategories (I.e.1, I.e.2, and III.b.2). The structure of the MET is as follows:

    I. ROADWAY SEGMENT
        I.a Segment Location Linkage
        I.b. Segment Roadway Classification
        I.c. Segment Cross Section
            I.c.1. Surface Descriptors
            I.c.2. Lane Descriptors
            I.c.3. Shoulder Descriptors
            I.c.4. Median Descriptors
        I.d. Segment Roadside Descriptors
        I.e. Other Segment Descriptors

      I.e.1. Bridge Descriptors

      I.e.2. Railroad Crossing Descriptors

    I.f. Segment Traffic Flow Data

    I.g. Segment Traffic Operations / Control Data

  II. Roadway Alignment Descriptors

    II.a. Horizontal Curve Data

    II.b. Vertical Grade Data

  III. Roadway Junction Descriptors

    III.a. At-Grade Intersection /Junctions

      III.a.1. General Descriptors

      III.a.2. At Grade Intersection /Junction Descriptors-Each Approach

    III.b. Interchange and Ramp Descriptors

      III.b.1. General Interchange Descriptors

      III.b.2. General Ramp Descriptors

The MET is implemented as a database (see Appendix A) on an FHWA server[3] and contains several data fields as described in Table 2. The mapping of elements to the three primary source documents is preserved in the following three MET data fields: MMIRE Corresponding Elements, SHRP-2 Corresponding Elements, and NCDOT Corresponding Elements.

---

[3] For information on how to access the database contact Lincoln Cobb, FHWA Office of Safety R&D, HRDS-2, (Lincoln.Cobb@dot.gov).

Table 2. Description of the Data Fields in the MET

Field Name	Description
DerivedFrom	List of element IDs from which this element is derived
ElementDescription	Description of the element as given in the MMIRE, SHRP, or NCDOT
ElementID	Number of the data element
ElementName	Name of the data element
HPMS	Data element required in: U = Universe file, S = Sample section Note: This categorization was only done for the MMIRE elements
IHSDM	Data element required: Y = Yes, O = Optional Note: This categorization was only done for the MMIRE elements
IsDerived	This column indicates whether the data element can be derived from the knowledge obtained for other data elements. Example: "Number of Exclusive Right Turn Lanes" could be derived from knowing the number of lanes, the number of thru lanes, and the number of exclusive left turn lanes.
IsNonVehicle	This column indicates whether the data element can be collected from a moving vehicle or not. Example: "Crossing pedestrian count/exposure" could not be collected from a moving vehicle.
MMIRECategory	1st level MMIRE classification
MMIRECorrespondingElements	Corresponding element in the MMIRE list Note: This column along with Columns X and Y provides a mapping of the MET elements back to the MMIRE, SHRP, and NCDOT elements.
MMIREEaseofDataCollection	"Estimate of the ease of data collection for that variable—easy (E), moderately difficult (M), difficult (D)" as given in the MMIRE.
MMIREPriority	Priority as given in the MMIRE: 1st priority or 2nd priority
MMIRESubCategory	2nd level MMIRE classification
MMIRESubSubCategory	3rd level MMIRE classification
MMUCC	This column indicates where this data element is part of the MMUCC. Note: This categorization was only done for the MMIRE elements
NCDOTCorrespondingElements	Correspondence of the data element to the NCDOT element. Note: This column along with Columns P and X provides a mapping of the MET elements back to the MMIRE, SHRP, and NCDOT elements.
RequiredMeasurementUncertainties	Where available, measurement uncertainties for the data element as required by various governmental agencies, standards, etc.
SafetyAnalyst	Data element required: M = Mandatory, O = Optional, S = Supplemental Note: This categorization was only done for the MMIRE elements
SameAsElements	This column indicates whether the data element is the same as another element. For example, the technology to identify "right shoulder type" would be the same as the technology to identify "left shoulder type."
Scope	Whether the data element will be analyzed as part of this project: In / Out
SHRP2CorrespondingElements	Correspondence of the data element to the SHRP-2 element. Note: This column along with Columns P and Y provides a mapping of the MET elements back to the MMIRE, SHRP-2, and NCDOT elements.
TechGap	Preliminary gap assessment based on number of technologies found. "Addressed" means 5 or more COTS technologies were found; "Partial" means between 1 and 4 COTS technologies were found; and "Not" means no COTS technologies were found.
TechnologyIDs	List of technologies that would address the data element
TechsAll	Number of technologies found
TechsCOTS	Number of COTS technologies found
TSIMS	Data element required: M = Minimum, B = Basic, E = Extended Note: This categorization was only done for the MMIRE elements

3. Technology Table

A literature review was conducted for each element in the MET to identify relevant technologies, to perform an assessment of the readiness of those technologies (where information was available), and to document providers and researchers involved with the technology delivery. Searches were performed for elements determined to be within the scope of the examination. Scope was defined primarily as measurable from a moving platform (either terrestrial or airborne) and not derived from another source (e.g., an existing roadway database or through measurement of other elements within the MET). Technologies that were identified were tagged with the following meta-data:

- What elements would this technology serve?
- What is the estimated technology readiness? (e.g., COTS, R&D, Prototype/Demonstration, Concept).
- What is the data processing speed? (e.g., real-time, offline post-processing).
- What is the data processing automation level? (e.g., automated, semi-automated, manual).
- What general category of technology does the specific technology incorporate? (e.g., the technology 'uses -' laser-based 3D imaging, camera-based 3D imaging, GPS, inertial systems, software only (mainly for research algorithms), laser-based 2D imaging, camera-based 2D imaging, or other).
- What general functional capability does the specific technology provide? (e.g., the technology 'provides -' location measurement, object or feature recognition, sign-type recognition, character recognition, spatial measurement, material optical property measurement, or material property measurement).

It was difficult to extract technology readiness level, data processing speed, and data processing automation level from information available during the literature review/market analysis. Most commercial vendors do not openly discuss this type of information.

As stated, the 392 data elements in the MET were filtered to sub-select the elements for which technology searches were conducted. Table 2 shows fields which indicate elements considered "Not in Scope" (Field Name = Scope), "Derived" (Field Name = IsDerived), "Not Measured from a Moving Vehicle" (Field Name = IsNonVehicle), or "Same as another element" (Field Name = SameAsElement). Technology searches were not conducted for those elements.

A Technology Table was created which contains a list of technologies that address the data elements in the MET. This list of technologies was compiled by conducting a search based on the data elements. The method used to conduct the search is given in Appendix B. The Technology Table is implemented as a database (see Appendix A) on a FHWA server and contains several data fields as described in Table 3.

Table 3. Description of the Data Fields in the Technology Table

Field Name	Description
AddressedElementIDs	List of elements addressed by this technology
ContactInformation	Contact information for the technology
DataProcessingAutomationLevel	Is the data processing: Automated; Semi-automated; Manual; Unknown
DataProcessingSpeed	Is the data processed: Real time; Offline; Unknown
Description	Description of the technology. A Brief Description field is available in the summary view on the SharePoint server and contains the first 200 characters from the "Description" field.
MeasurementUncertainties	Uncertainties of the measurements as given by the proprietor
OtherInformation	Other information about the technology
PotentialElementComments	Comments for how this technology could address the potential elements
PotentialElementIDs	List of elements potentially addressed by this technology
ProductName	Name of the product
ProprietorName	Proprietor or owner of the technology
ProvidesCharacterRecognition	Does the technology recognize characters in a sign, that is, can it read signs?
ProvidesLocationMeasurement	Does the technology provide location measurements
ProvidesMaterialPropertyMeasurement	Does the technology provide information about the material of the object, i.e., is it wood, metal, concrete, metal, paint, polyurea?
ProvidesMaterialPropertyOpticalMeasurement	Does the technology provide the optical properties of an object such as the color and reflectivity?
ProvidesObjectOrFeatureRecognition	Does the technology provide object or feature recognition such as crack type recognition and traffic signal detection?
ProvidesSignTypeRecognition	Does the technology recognize signs and the type of signs?
ProvidesSpatialMeasurement	Does the technology provide spatial measurements such as the location (x, y, z) of an object or the dimensions of objects?
TechID	Technology identification number
TechReadiness	Level of readiness of the technology: COTS - Commercial-off-the-Shelf; Prototype or Demo; Research and Development; Concept
UsesAirborneSystem	Is the technology airborne-based?
UsesCameraBased2DImaging	Does the technology use camera-based 2D imaging such as video or camera?
UsesCameraBased3DImaging	Does the technology use camera-based 3D imaging systems such as stereo video or photogrammetry?
UsesGPS	Does the technology use GPS?
UsesInertialSystem	Does the technology use inertial navigation systems such as gyroscopes and accelerometers?
UsesLaserBased2DImaging	Does the technology use laser-based 2D imaging such as a laser profilometer or a line scanner?
UsesLaserBased3DImaging	Does the technology use laser-based 3D imaging systems such as laser scanners?
UsesOther	Does the technology use some other technology such as spectrometer?
UsesSoftwareOnly	Is the technology an algorithm?
UsesTerrestrialSystem	Is the technology terrestrial-based?

The categories "Description of the Technology" and "Measurement Uncertainties" contain descriptions and uncertainties, which in most cases, were obtained from the technology proprietor's literature. No attempt was made to verify those claims.

For a given technology, the "Data Processing Speed" and the "Data Processing Automation Level" were rated based upon best judgment. That is, this information is often not indicated in the literature or it was unclear if the process was performed in real-time or automatically. Also, the claims of "automated" cannot be verified or assessed without a much more in-depth review of the applicable technology (e.g., system-level testing). In cases when there was no indication or if it was not clear, the entry made for these columns was "Unknown" (conservative). When the technologies perform several functions, some automated and some not, the entry made was semi-automated. The data in the Technology Table is shown graphically in Figures 3.1 to 3.6.

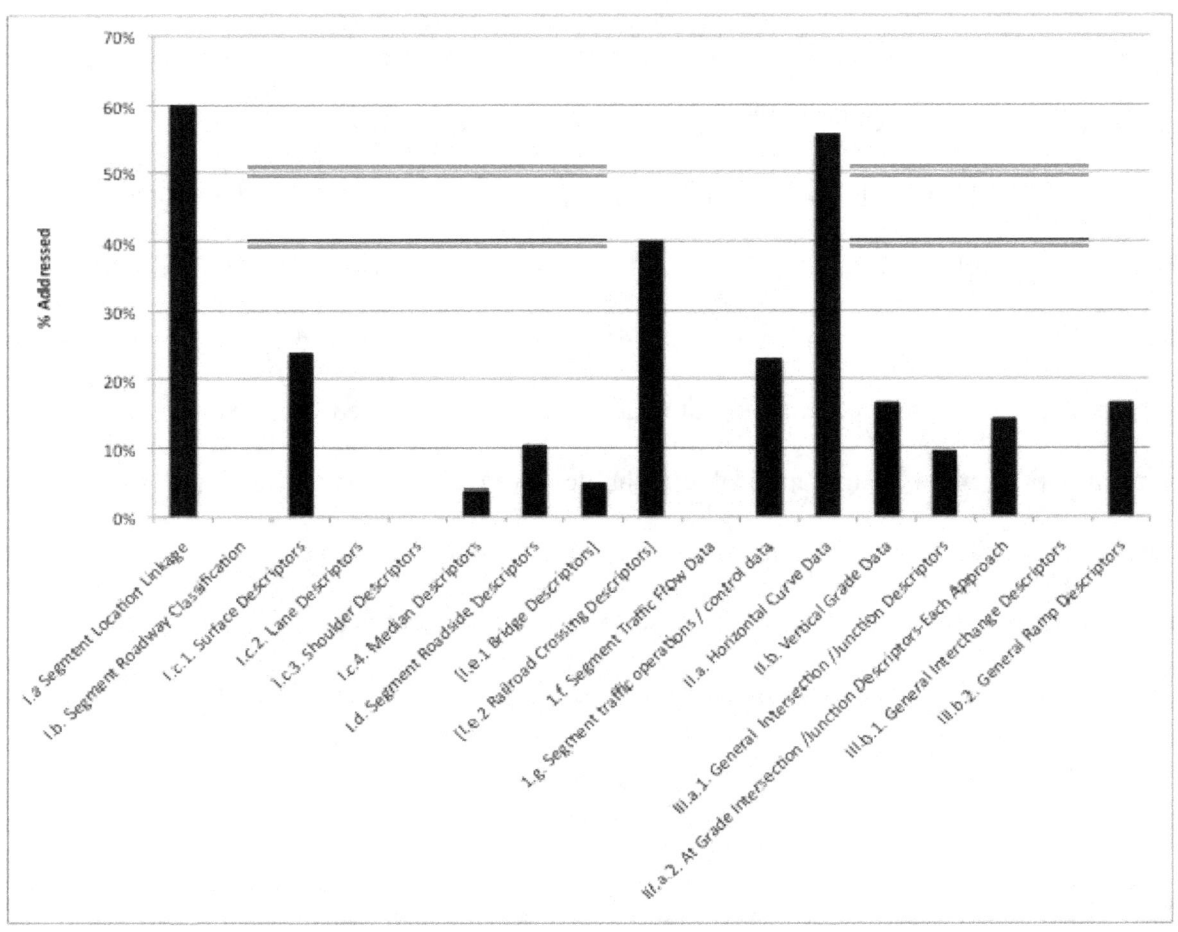

Figure 3.1. Percentage of Addressed elements within each MET category.

Figure 3.1 shows a column chart of the percentage of elements within each MET category for which five or more commercial technologies (i.e., "Addressed" elements) were identified. This chart gives an indication of the maturity, in terms of commercialization, of the technologies within each of the MET categories.

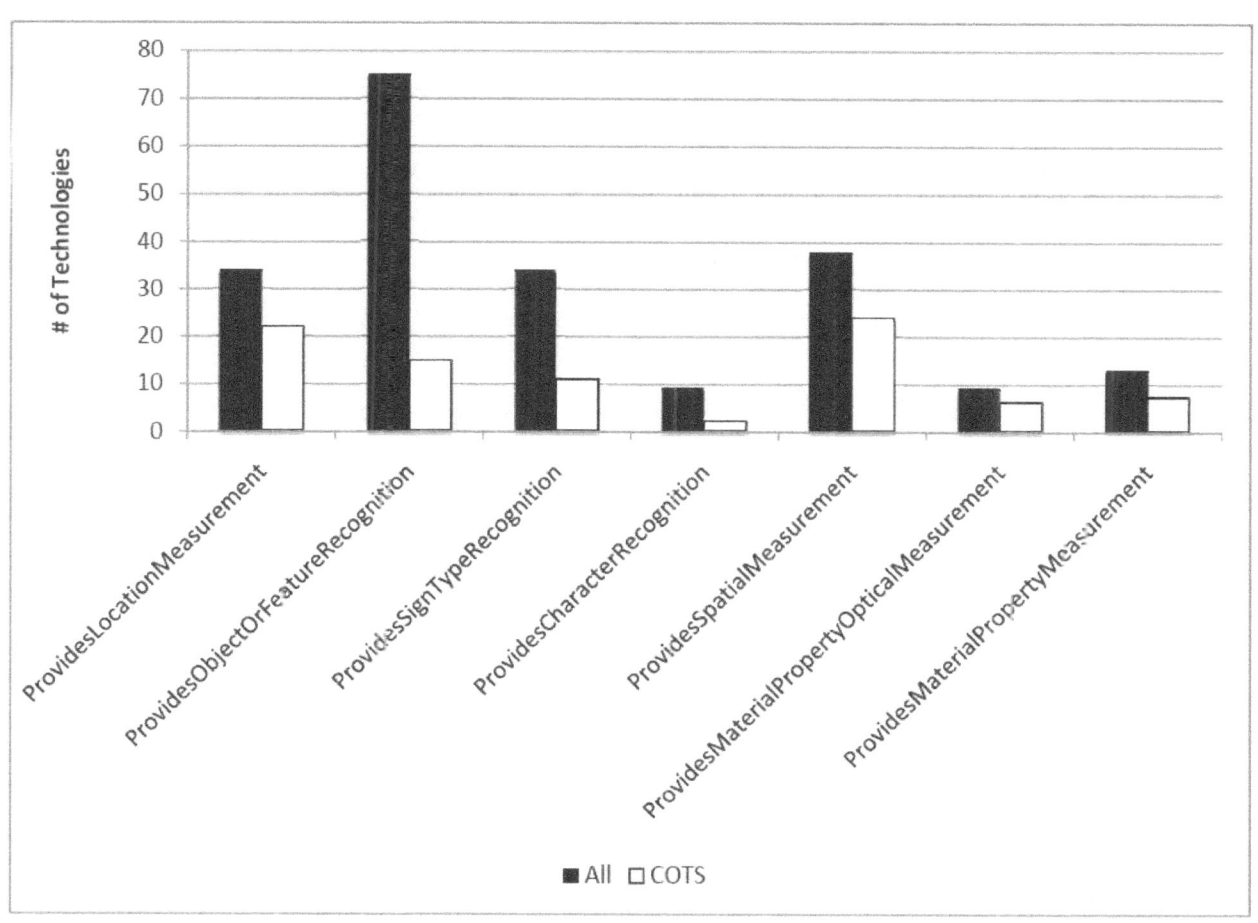

Figure 3.2. Number of technologies in each of the technology functional classification categories.

Figure 3.2 shows the number of technologies in each technology functional classification category. The black columns show the total number of technologies in each category while the white columns show the number of commercial technologies in each category.

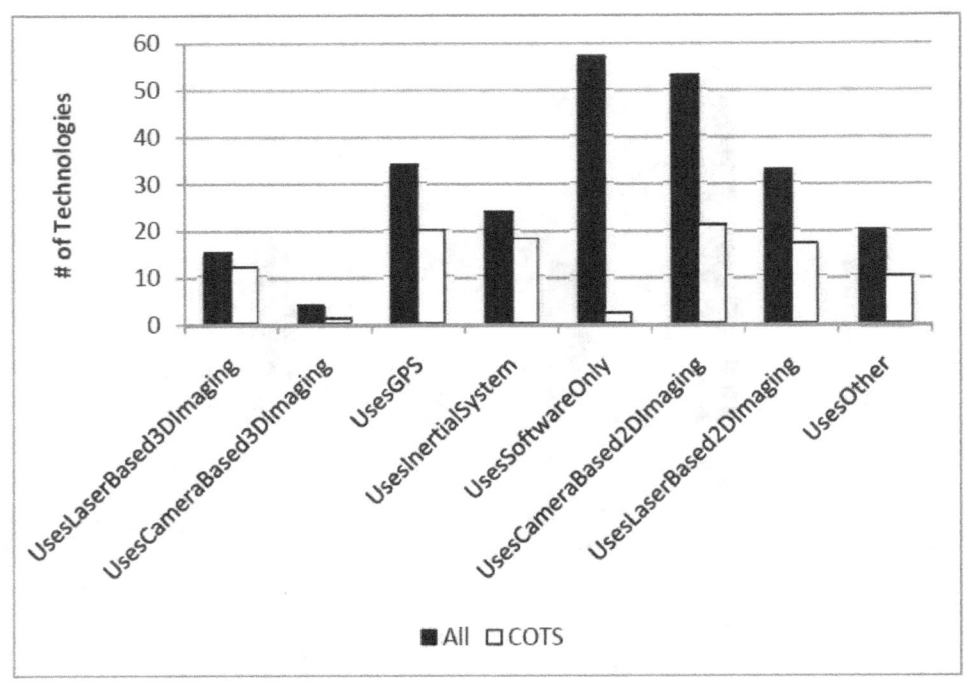

Figure 3.3. Number of identified technologies grouped by general technology category.

Figure 3.3 shows the number of technologies within each general technology category for all technologies (black) and for only the commercial technologies (white).

Figure 3.4. Number of elements addressed (black) and potentially addressed (white) by technologies T001 to T072.

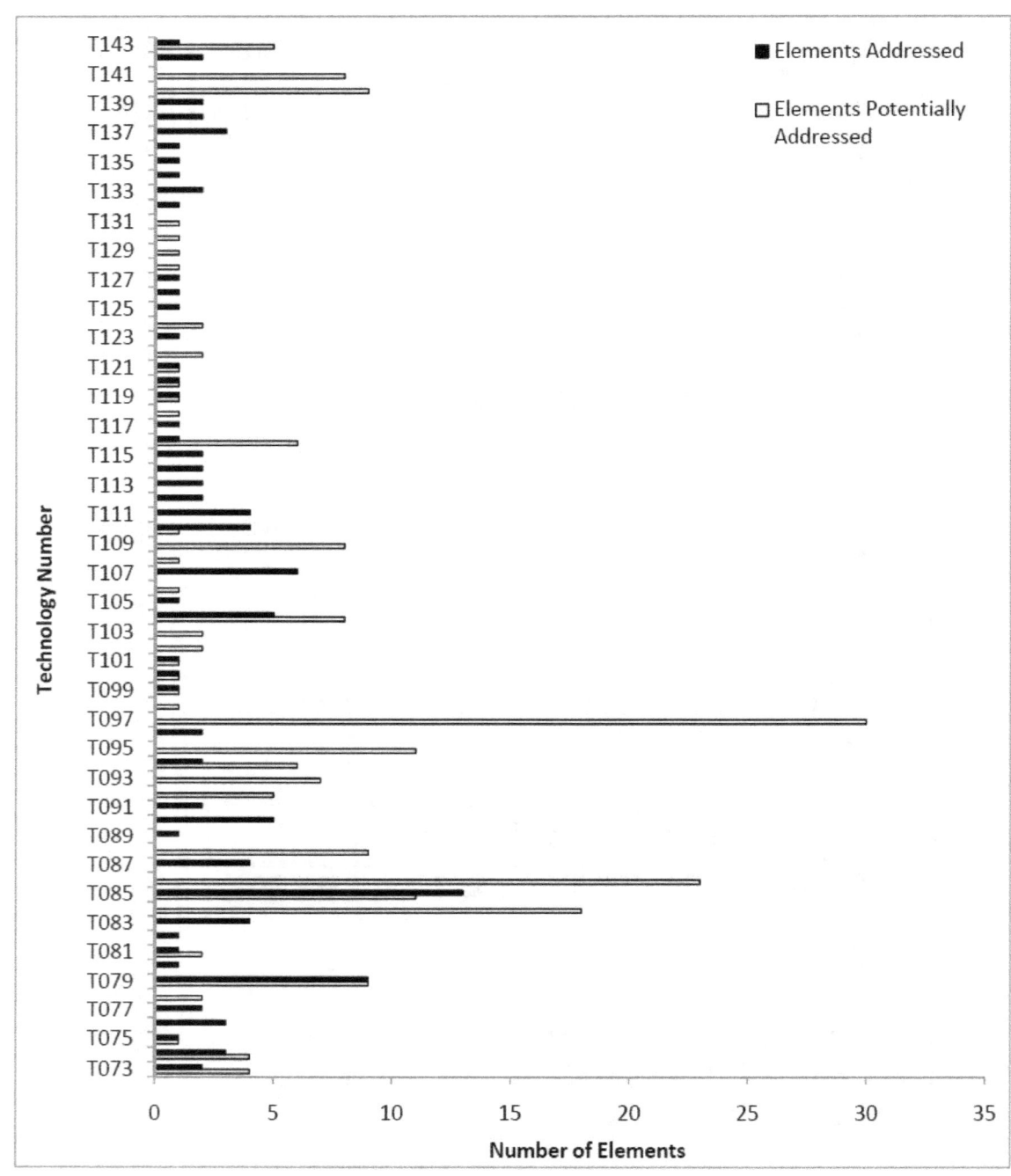

Figure 3.5. Number of elements addressed (black) and potentially addressed (white) by technologies T073 to T143.

Figure 3.4 and Figure 3.5 show the number of elements addressed (black) and potentially addressed (white) by each of the technologies identified in this study. These figures give an indication of the number of elements that are impacted by or could potentially be impacted by a technology.

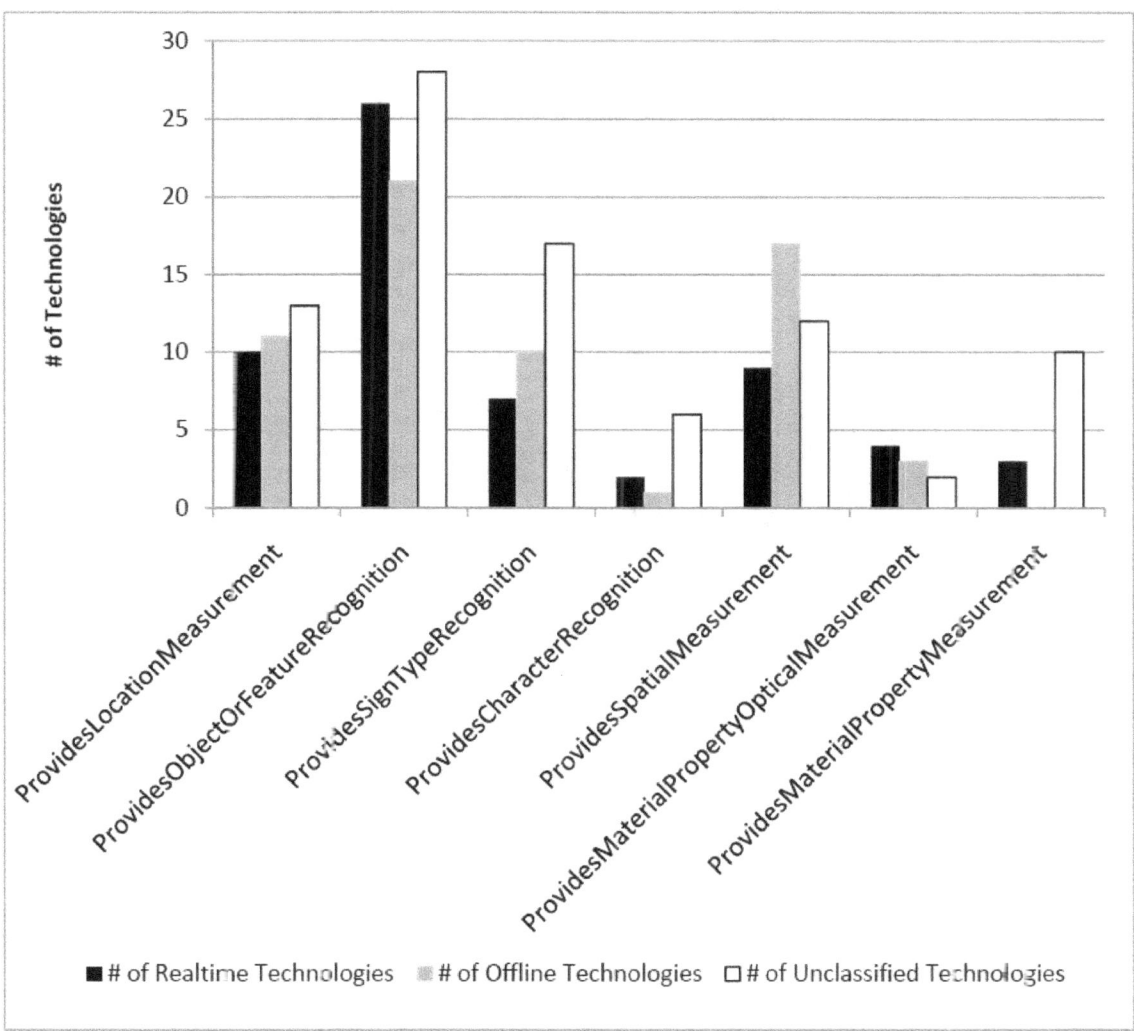

Figure 3.6. Number of technologies within each functional classification category that provide real-time, offline, and unknown data processing speed.

Figure 3.6 shows a column chart of the number of technologies within each functional classification category in which the technologies are divided into three groups based on their data processing speed (real-time, offline, and unclassified).

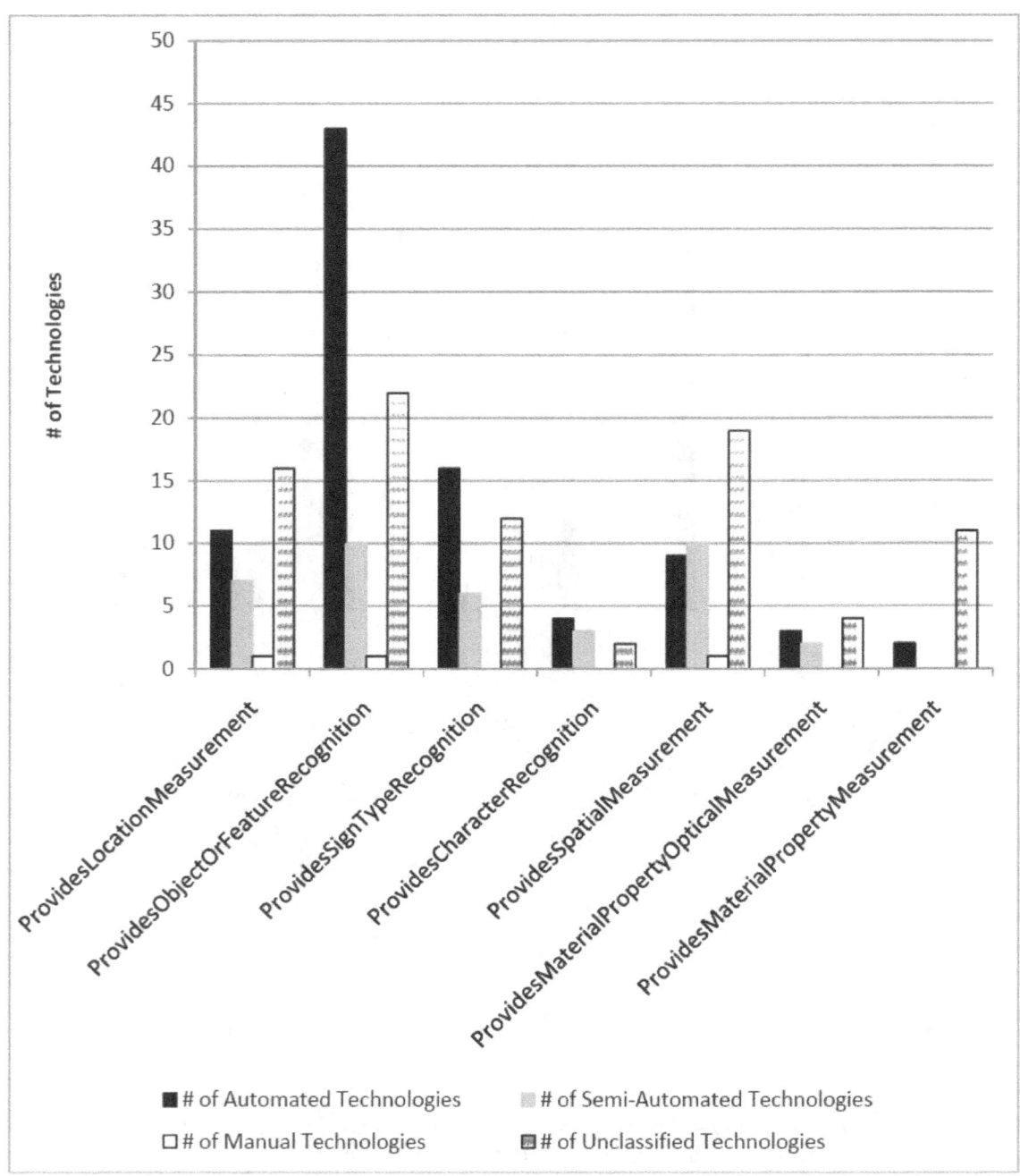

Figure 3.7. Number of technologies within each functional classification category that provide automated, semi-automated, manual, and unknown levels of data processing automation.

Figure 3.7 shows a column chart of the number of technologies within each functional classification category in which the technologies are divided into four groups based on their level of data processing automation (automated, semi-automated, manual, and unclassified). Of note, the differentiation between automated and semi-automated is subjective, and often based upon incomplete data.

4. Technology Gap Assessment

This section presents the procedures used to assess the technology gaps based on the concept of demand and maturity. The demand vs. maturity graph gives an indication of how well the technology demands are addressed by both the market and the research community. The general concept of the demand vs. maturity graph is shown in Figure 4.1.

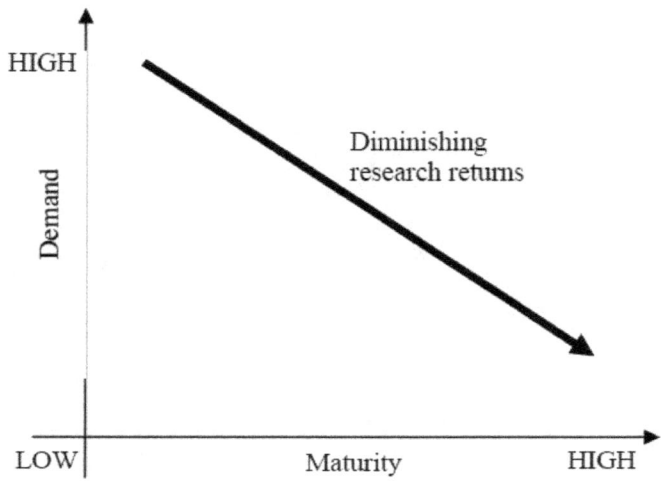

Figure 4.1. Schematic of the Demand vs. Maturity Graph.

Some preliminary steps prior to the assessment include definitions of the terms and variables that will be used in the analyses. These definitions are given in Section 4.1. The technologies were grouped by functions (i.e., what function does the group of technologies provide?). These groupings and their explanations are provided in Section 4.2. The technology gaps were determined for these Technology Functional Classification Categories. The framework for the analysis is shown in Figure 4.2.

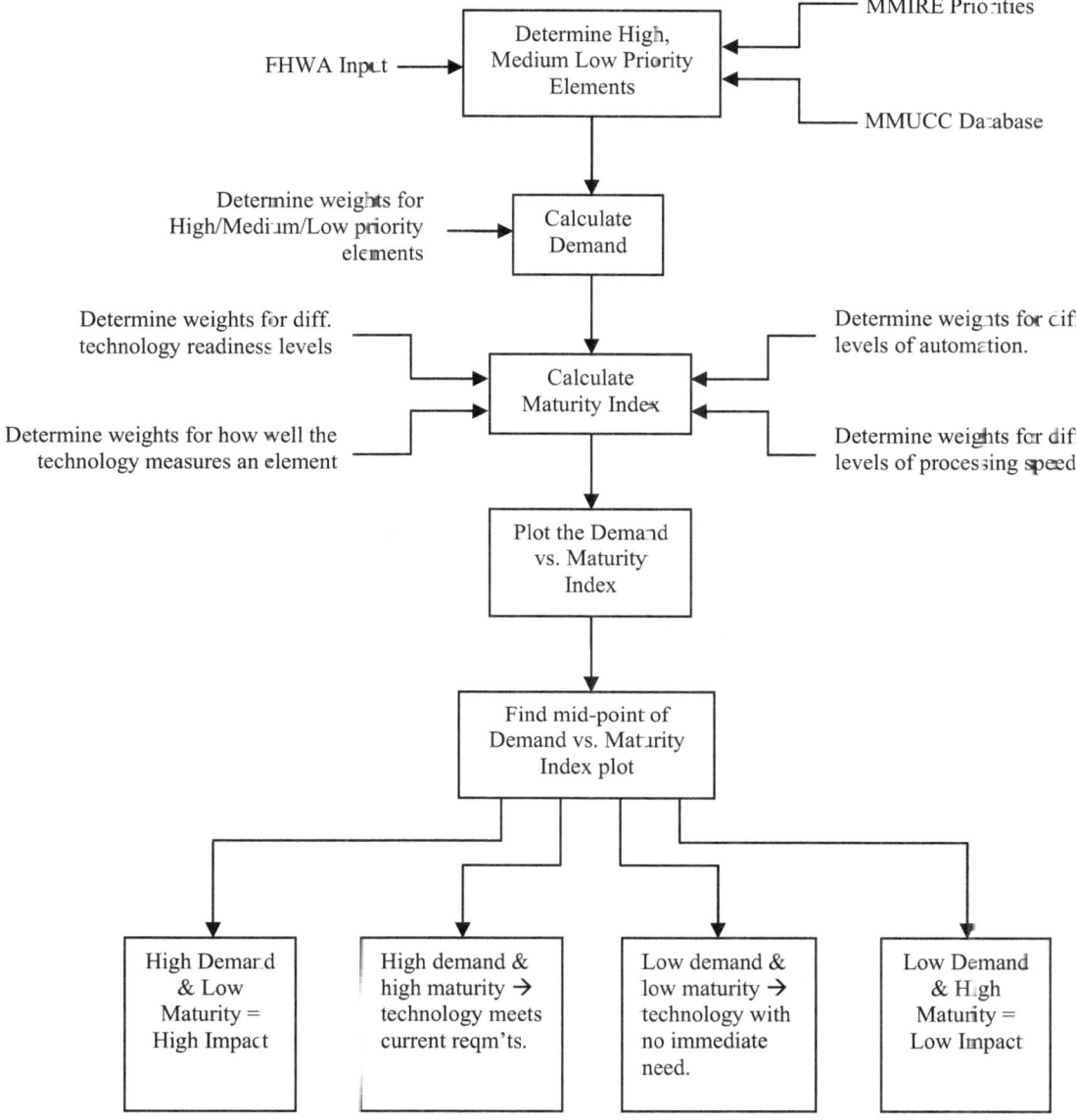

Figure 4.2. Framework for the analysis procedure.

4.1 Definitions

4.1.1 Demand

Demand is related to the number of elements that require a measurement capability provided by a particular technology functional classification category. It is defined as the weighted sum of high, medium, and low priority elements within a technology functional classification category divided by the weighted sum of the total number of elements in-scope:

$$D_i = \frac{w_{high} * E\_High_i + w_{med} * E\_Medium_i + w_{low} * E\_Low_i}{w_{high} * E\_Hi_{total} + w_{med} * E\_Med_{total} + w_{low} * E\_Low_{total}} \quad Eq.\,1$$

where

D_i	=	the demand for technology functional classification category i. The value lies between [0, 1].
E_High_i	=	the number of in-scope high priority elements in functional classification category i
w_{high}	=	the weight for high priority elements
E_Medium_i	=	the number of in-scope medium priority elements in functional classification category i
w_{medium}	=	the weight for medium priority elements
E_Low_i	=	number of in-scope low priority elements in functional classification category i
w_{low}	=	the weight for low priority elements
E_Hi_{total}	=	the total number of in-scope high priority elements
E_Med_{total}	=	the total number of in-scope medium priority elements
E_Low_{total}	=	the total number of in-scope low priority elements

4.1.2 Maturity Index

The Maturity Index is an indication of the current state of where a technology functional classification category is located along a linear scale that goes from immature (0) to mature (1). In this report, the maturity of a technology functional classification category depends on its technology readiness level, level of automation, data processing speed, and its ability to measure an element to the required accuracy or resolution. A fully mature technology functional classification category is one where all the technologies within that category are commercially available and fully automated and can process data in real-time and measure elements to the required levels of accuracy or resolution.

Therefore, the Maturity Index (MI) of a technology functional classification category is defined as the weighted sum of the maturity indices for the category's technology readiness level, level of automation, data processing speed, and the ability to measure an element to the required accuracy or resolution (see Eq. 2). A MI of 1 would indicate that the technology functional classification category is fully mature.

$$MI = w_1 * M_{TRL} + w_2 * M_{auto} + w_3 * M_{speed} + w_4 * M_{spec} + \cdots + w_n * M_n \qquad Eq.2$$

Where

w_1 = weight for the technology readiness level (TRL)

w_2 = weight for level of automation of technology

w_3 = weight for processing speed of technology

w_4 = weight for how well the technology meets the accuracy or resolution requirement for an element

w_n = weight for Maturity index n

M_{TRL} is the maturity index of the TRL of the functional category,

$$M_{TRL} = \frac{w_c*T_c + w_{rd}*T_{rd} + w_p*T_p + w_{cots}*T_{cots}}{T_c + T_{rd} + T_p + T_{cots}}$$

where

$w_c, w_{rd}, w_p, w_{cots}$ = weights for the technology readiness subcategories Concept, R&D, Prototype/Demo, COTS, respectively. See Section 4.1.4 for definitions of these subcategories.

T_c, T_{rd}, T_p, T_{cots} = number of technologies within the technology functional classification categories that are Concept, R&D, Prototype, and COTS, respectively.

M_{auto} is the maturity index of the level of automation of the functional category,

$$M_{auto} = \frac{w_{auto}*A_{auto}+w_{semi}*A_{semi}+w_{manual}*A_{manual}}{A_{auto}+A_{semi}+A_{manual}}$$

Where

w_{auto}, w_{semi}, w_{manual} = weights for the levels of automation fully automated, semi-automated, and manual, respectively.

T_{auto}, T_{semi}, T_{manual} = number of technologies within a technology functional classification category whose level of automation is fully automated, semi-automated, and manual, respectively.

M_{speed} is the maturity index of the data processing speed of the functional category,

$$M_{speed} = \frac{w_{realtime}*S_{realtime}+w_{offline}*S_{offline}}{S_{realtime}+S_{offline}}$$

Where

$w_{real-time}$, $w_{offline}$ = weights for the processing speeds real-time and offline, respectively.

$T_{real-time}$, $T_{offline}$ = number of technologies within a technology functional classification category whose processing speed is real-time and offline, respectively.

$w_{realtime}, w_{offline} =$

M_{spec} is the maturity index of how well the technologies within a functional category meet the requirements of the elements within the same functional category,

$$M_{reqmt} = \frac{w_{meet}*R_{meet}+w_{partial}*R_{partial}+w_{min}*R_{min}}{R_{meet}+R_{partial}+R_{min}}$$

Where

$W_{exceed}, W_{meet}, W_{not\,meet}$ = weights for the ability to exceed, meet, not meet, respectively, the requirements of the elements within a functional category.

$T_{exceed}, T_{meet}, T_{not\,meet}$ = number of technologies within a technology functional classification that exceed, meet, do not meet, respectively, the requirements of the elements within the same functional category.

M_n is the maturity index n.

4.1.3. Impact

Impact is defined as regions in the Demand vs. Maturity plot that are centered at the midpoint of the actual Demand and Maturity values (see Figure 4.3).

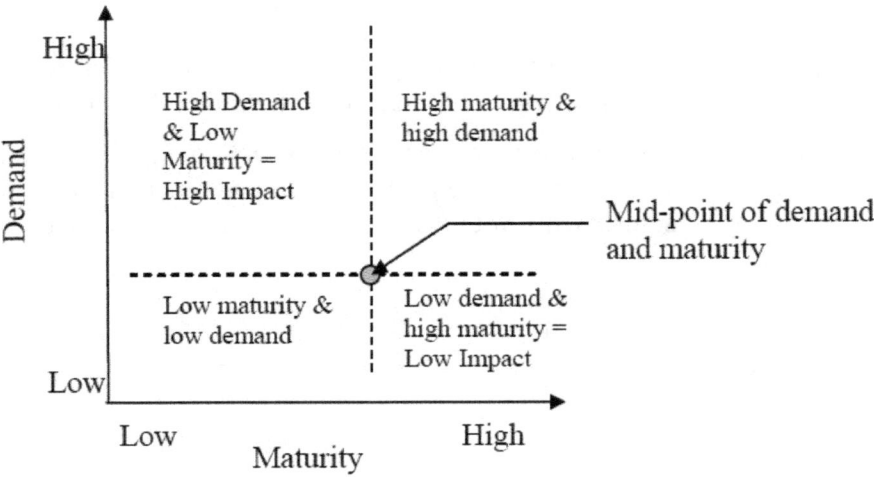

Figure 4.3. Impact as defined by the Demand vs. Maturity Plot.

The different regions of the Demand vs. Maturity plot represent different impacts or research priorities:

- High demand and low maturity indicates a **HIGH** impact region

- Low demand and high maturity indicates a **LOW** impact region

- High demand and high maturity indicates a technology that meets current requirements. Research in this area could be beneficial if the future performance need is expected to increase. Because of the high initial level of maturity, additional research funding will experience diminishing returns and require additional time to obtain the benefits. However, because many elements are affected by this technology, the investment is justifiable.

- Low demand and low maturity indicates a technology with no immediate need. Although the low initial maturity may result in rapid breakthroughs, the relatively few number of elements impacted by these developments may not justify an investment in this area. However, this area should be closely monitored as technological improvements could lead to new applications that address additional elements.

4.1.4. Technology Readiness Level Subcategories

The technology readiness level (TRL) is used to determine where a technology is in terms of commercialization where commercially available is the desired goal. The subcategories of TRL are defined as follows:

- **COTS** = Commercial-off-the-shelf. Technology is commercially available.
- **Prototype/Demonstration** = A technology is not yet commercially available. A prototype of the technology has been demonstrated under representative conditions.
- **R&D** = Research and Development. The technology is at the proof-of-concept stage. Hardware: Component or model validation in a laboratory environment. Software: Development of limited functionality to validate critical properties and predictions using non-integrated software.
- **Concept** = Idea or patent for a specific application is identified. No experimental data or detailed analysis is available to support the concept.

4.2. Technology Functional Classification Categories

4.2.1. Location Measurement

Location measurement refers to calculating a geographic position relative to a defined reference frame. This measurement can be an absolute position calculation in a fixed reference frame (e.g. GPS, WAAS), or a dead-reckoned position (e.g., INS) relative to a known benchmark. Multiple location measurement technologies can be combined to overcome technical limitations of a single system. The accuracy of GPS receivers is dependent on a number of parameters including the number of satellites in view, inherent satellite signal accuracy, signal transmission errors, and receiver hardware and software limitations. The accuracy of measurements can be improved by post-processing or applying differential corrections.

4.2.2. Object and Feature Recognition

Object and feature recognition refers to the ability to identify:

- Objects such as street lights, road signs, and barriers
- Conditions of identified objects (e.g., damaged barrier, blockage)
- Features such as the type of cracking

The ideal case would be to automatically extract the information from the data in a reliable manner using a robust algorithm without any user intervention and while passing the object or feature at highway speeds. The algorithm depends on the sensor(s) being used to collect the data and the object itself. Every recognition algorithm consists of two parts: 1) determining the numerical values of the parameters characterizing the object (e.g., aspect ratio of object, expected size), and 2) classifying whether the candidate object belongs to a given category.

Two common technologies used to obtain the data are 3D imaging systems[4] and video cameras. 3D imaging systems provide direct measurements of object position and dimensions. The manufacturer specified range errors for these systems are on the order of 4 mm to 8 mm[5]. As there are no standards for these systems, it is difficult to quantify level of detail that can be captured by these systems as this ability is affected by factors such as beam width, object reflectivity, distance to the object, angle of incidence with the object, object material/texture, and scan speed (for scanning systems).

---

[4] A non-contact measurement instrument used to produce a 3D representation (for example, a point cloud) of an object or a site. Some examples of a 3D imaging system are laser scanners (also known as LADARs or LIDARs or laser radars), optical range cameras (also known as flash LIDARs or 3D range cameras), triangulation-based systems such as those using pattern projectors or lasers, and other systems based on interferometry {ASTM, 2009, Standard Terminology}.

[5] There are currently no standards for quantifying the performance of 3D imaging systems. This need is being addressed by the ASTM E57 3D Imaging Systems standards committee.

4.2.3 Sign Type Recognition

Sign recognition is a special case of object recognition. Standard sign dimensions, shapes, and colors make this technology more robust than the generalized object recognition technologies described in Section 4.2.2. Common failure modes for sign recognition systems include off-angle observations, signs with debris or damage, non-standard or partially obscured signs, and environmental conditions.

4.2.4. Character Recognition

Character recognition in this study is limited to the ability to "read" signage (e.g., speed limits, parking restrictions, or railroad crossing numbers). Character recognition is necessary when sign type recognition fails, such as in cases where signs are non-standard (e.g., in toll facilities) or the meaning of the sign depends solely on the text (e.g., speed limits).

For roadside asset inventory, the data is often extracted from video images. Text recognition from video images is much more difficult than text recognition (optical character recognition or OCR) from scanned images. However, unlike commercial OCR software, text recognition for roadside asset inventory benefits from reduced complexity due to standardized fonts, font sizes, and a limited number of pre-defined text strings. The field of video OCR appears to be fairly mature and is growing as there is an increasing need to search for text in video in other sectors.

4.2.5. Spatial Measurement

Spatial measurement refers to obtaining the 3D coordinates of roadway structures. In contrast to location measurement, positions are derived from photogrammetric analysis or 3D point cloud measurements rather than the absolute position provided by GPS. Spatial measurement also extends to determining dimensions of structures (e.g., length of a turn lane, road profile).

The two common technologies to obtain spatial measurements are 3D imaging systems and video cameras. See Section 4.2.2 for a discussion of these technologies. Another common technology is the use of a laser line scanner to obtain the road profile.

4.2.6. Material Optical Property Measurement

Material optical property measurement refers to the reflectivity and color of signs and lane markings. Potential technologies for this measurement include cameras and retroreflectometers. There are standard test methods for the measurement of the retroreflective properties of pavement markings (ASTM E1710) and signs (ASTM E1709). There is a federal rule which went into effect on Jan. 22, 2008 on maintaining traffic sign retroreflectivity. "The purpose of this final rule is to revise standards, guidance, options, and supporting information relating to maintaining minimum levels of retroreflectivity for traffic signs on all roads open to public travel."[6] This federal rule will likely spur innovations towards more accurate and real-time measurements in this area.

4.2.7. Material Property Measurement

Material property measurement considers the type of material (such as concrete or asphalt, wood or metal) as well as material surface characteristics (e.g., friction, texture). Road surface materials are particularly relevant, but other structures are applicable as well, including

---

[6] Federal Register, Vol. 72, No. 245, Dec. 21, 2007, Rules and Regulations, p. 72574-72582.

lane markings and posts. Very few automated technologies for determining material type have been identified. There are COTS technologies available to measure roadway friction and texture and standards are available on how to obtain these measurements. There has been some work to determine if a roadway was asphalt or concrete based on analysis of the reflection and absorption characteristics of a light source on different surfaces (spectral analysis).

4.3 Determining High, Medium and Low Priority Elements

High priority elements were identified as those derived from the following:

- Survey of subject matter experts
- FHWA primary focus areas[7]
- MMIRE designated high-priority elements (those with priority 1)
- MMUCC database[8]
- Communication with FHWA[9]

A list of 100 high priority elements (see Table C.1 in Appendix C) was created based on the MMIRE priority, MMUCC database, FHWA primary focus areas, and communication with FHWA (elements not on this list were considered low priority). This list was then sent to subject matter experts in FHWA for further refinement in terms of priority and whether they were addressed or not. The survey responses are presented in Table C.2 in Appendix C. Based on the survey results, the 100 elements were reclassified as high, medium, or low priority. The reclassification was based on the number of "Agree" vs. the total number of "Agree" and

---

[7] FHWA primary focus areas are intersections, pedestrian and bicyclist safety, roadside safety, run-off-road safety, speed management (FHWA, 2009b). Intersection and intersection-related crashes consistently make up a high proportion of total fatal crashes, up to 23 %, more than 50 % of the combined fatal and injury crashes occur at intersections (FHWA, 2009a)
[8] See http://www.mmucc.us/dataelements/roadway-intro.aspx
[9] Conversation with FHWA (L. Cobb, C. Tan, R. Pollack on May 12, 2009)

"Disagree"[10] (where R is the ratio of the number of "Agree" responses to the sum of the number of "Agree" and "Disagree" responses). The following criteria were then used to reclassify the elements (R values were based on best judgment):

- High Priority: $R \geq 2/3$
- Medium Priority: $1/3 < R < 2/3$
- Low Priority: $R \leq 1/3$

The new list of high, medium, and low priority elements was then combined with a list of high, medium, low priority elements provided by FHWA (see Table D.1 in Appendix D). If the priority classification of an element was different in the two lists, the higher priority classification took precedence. The resulting prioritized list of elements is presented in Appendix E.

4.4. Demand Calculation

To calculate the demand, the weights used for the high, medium, and low elements were $w_{high} = 1$, $w_{med} = 0.5$, and $w_{low} = 0.3$, respectively. The weights were assigned based on discussions with FHWA. The numbers of high, medium, and low priority elements for each of the technology functional classification category are given in Table 4 and shown graphically in Figure 4.4. The demands for the technology functional classification categories were calculated using Eq. 1 and are also given in Table 4.

---

[10] Since there were no "Strongly Disagree" responses, that category was not used in the reclassification. Responses of "Don't Know" were also not used in the reclassification because we assumed that the respondents had no opinion on those elements.

Table 4. Number of High, Medium and Low Priority Elements and Demands

# of Elements	Location measurement	Object or feature recognition	Sign type recognition	Character recognition	Spatial measurement	Material optical property measurement	Material property measurement
High Priority	26	62	7	2	38	1	5
Medium Priority	7	17	8	2	8	0	3
Low priority	27	84	16	3	57	4	6
Demand	0.17	0.43	0.07	0.02	0.27	0.01	0.04

Demand can be represented by a value between 0 and 1 where either no or all elements are affected by a particular functional category, respectively. The values of 0 and 1 are theoretical constructs and are unlikely to be achieved in practice. In our analysis, typical values were between 0 and 0.5. Note that due to normalization, the sum of the demand for all categories will equal 1.0.

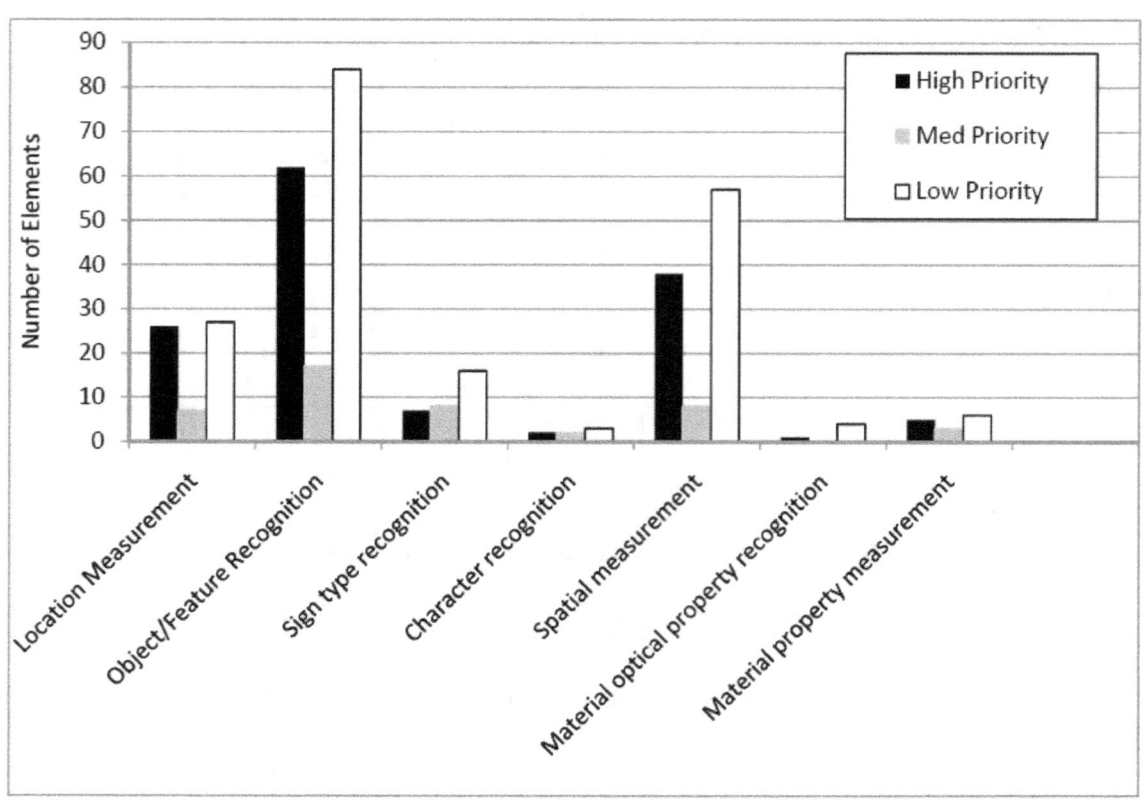

Figure 4.4. Number of elements belonging to each technology functional classification category.

4.5 Maturity Index Calculation

As shown in Eq. 2, the Maturity Index consists of several components. For this report, the Maturity Index is based only on the technology readiness level since the data for this aspect of the technology has the least uncertainty associated with it. The data collected for automation, data processing speeds and accuracy/resolution achievable by a technology could not be collected with sufficient detail or the information was not publicly available; thus, the classification of the subcategories for these Technology Functional Classifications in the Technology Table is subjective. Therefore, in this report, $w_1 = 1$ and w_2, w_3, $w_4 = 0$ in Eq. 2 until there are more objective means of determining levels of automation, processing speeds, and the levels of accuracy/resolution achievable by the technology.

The weights for the subcategories (Concept, R&D, Prototype/Demonstration, COTS) of the technology readiness level were based on an estimate of the probability that a particular subcategory will become COTS in 3 to 5 years. Based on discussions with FHWA, the weights were $w_c = 0.1$, $w_{rd} = 0.3$, $w_p = 0.7$ and $w_{cots} = 1.0$ for Concept, R&D, Prototype/Demo, and COTS, respectively. That is, there is a 70 % chance that a prototype would become COTS in 3 to 5 years. The calculation of the Maturity Index is shown in Table 5. In Table 5, the Sum of COTS Equivalents is the numerator of the M_{TRL} equation and the Maturity Index obtained by dividing this value by N (row 2 in Table 5).

Table 5. Maturity Index Calculation

		Location Measurement	Object / Feature Recognition	Sign Type Recognition	Character Recognition	Spatial Measurement	Material Property Optical Measurement	Material Property
N = Total # of technologies in the technology functional classification category		34	75	34	9	38	9	13
Concept ($w_c = 0.1$)	No. technologies (T_c)	0	4	1	1	0	0	0
	COTS equivalents = $w_c * T_c$	0	0.4	0.1	0.1	0	0	0
R&D ($w_{rd} = 0.3$)	No. technologies (T_{rd})	9	50	21	6	10	3	6
	COTS equivalents = $w_{rd} * T_{rd}$	2.7	15	6.3	1.8	3	0.9	1.8
Prototype ($w_p = 0.7$)	No. technologies (T_p)	3	6	1	0	4	0	0
	COTS equivalents = $w_p * T_p$	2.1	4.2	0.7	0	2.8	0	0
COTS ($w_{cots} = 1.0$)	No. technologies (T_{cots})	22	15	11	2	24	6	7
	COTS equivalents = $w_{cots} * T_{cots}$	22	15	11	2	24	6	7
	Sum of COTS equivalents	26.8	34.6	18.1	3.9	29.8	6.9	8.8
	MATURITY INDEX (MI)	**0.79**	**0.46**	**0.53**	**0.43**	**0.78**	**0.77**	**0.68**

An estimate of the sensitivity of the Maturity Index is $\frac{X}{(X+N)}$ where X is the number of additional COTS technologies found and N is as defined in Table 5. This value is the approximate change in the maturity index if additional COTS technologies were found in the literature search. If the additional technology found were other than COTS, the change would be less because the weight for each non-COTS technologies is less than the weight for COTS technologies. Table 6 lists approximate changes to the Maturity Index if additional COTS technologies were found.

Table 6. Sensitivity of Maturity Index to Additional COTS Technologies Found

		Location Measurement	Object / Feature Recognition	Sign Type Recognition	Character Recognition	Spatial Measurement	Material Property Optical Measurement	Material Property Measurement
N = Total # of technologies in the technology functional classification category		34	75	34	9	38	9	13
Change to Maturity Index = $X/(N+X)$	X = + 1 COTS found	0.03	0.01	0.03	0.10	0.03	0.10	0.07
	+ 2	0.06	0.03	0.06	0.18	0.05	0.18	0.13
	+ 3	0.08	0.04	0.08	0.25	0.07	0.25	0.19
	+ 4	0.11	0.05	0.11	0.31	0.10	0.31	0.24
	+ 5	0.13	0.06	0.13	0.36	0.12	0.36	0.28
	+ 6	0.15	0.07	0.15	0.40	0.14	0.40	0.32
	+ 7	0.17	0.09	0.17	0.44	0.16	0.44	0.35
	+ 8	0.19	0.10	0.19	0.47	0.17	0.47	0.38
	+ 9	0.21	0.11	0.21	0.50	0.19	0.50	0.41
	+ 10	0.23	0.12	0.23	0.53	0.21	0.53	0.43

4.6. Results

Figure 4.5 shows the demand versus maturity graph derived from values calculated in Sections 4.4 and 4.5.

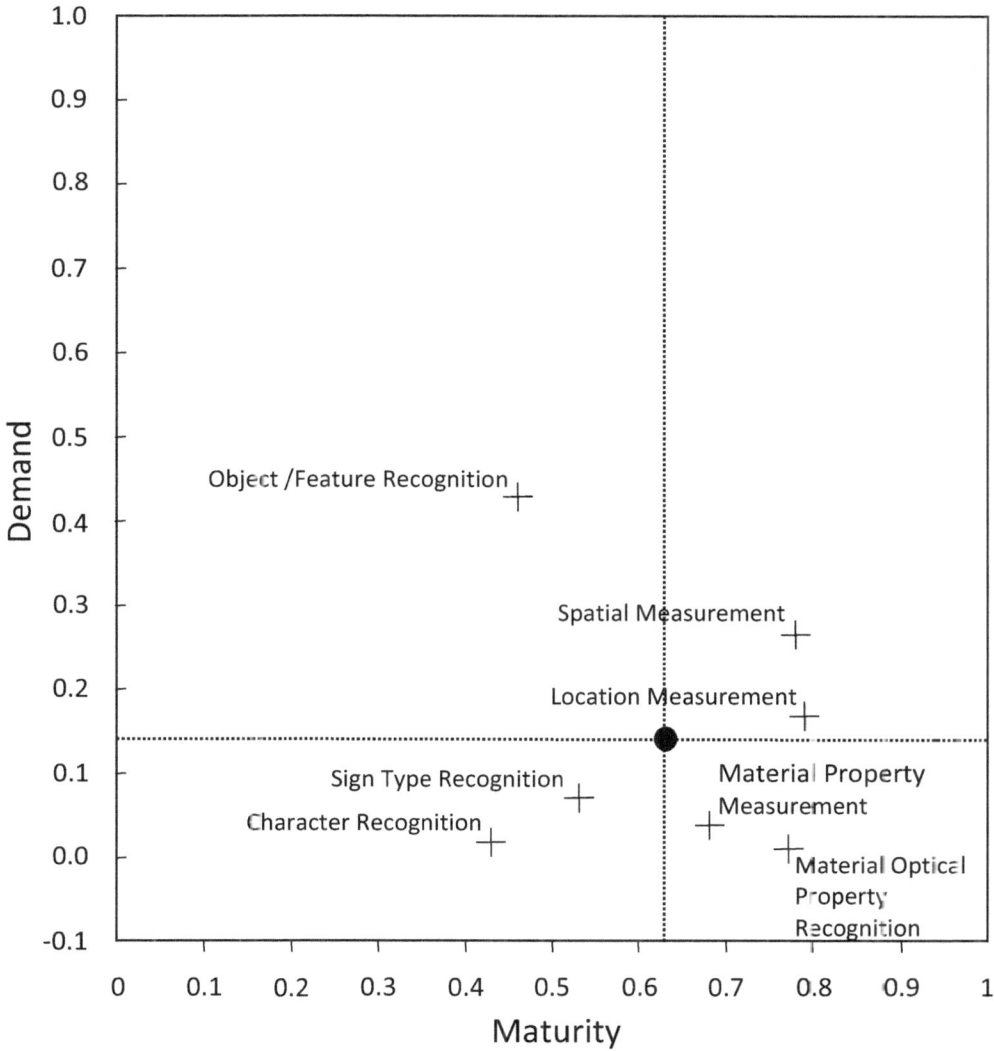

Figure 4.5. Demand vs. Maturity Graph.

From Figure 4.3 and Figure 4.5, it follows that (1) the Object/Feature Recognition technology functional classification category is in the high impact region, and (2) the Material Property Optical Measurement and the Material Property Measurement technology functional classification categories are in the low impact region.

In assessing the relative priorities of each functional classification category, the following observations are made:

- Object/feature recognition should be given highest priority.
- Material property optical measurement should be given the lowest priority.
- Spatial measurement should be given secondary priority in anticipation of increased accuracy requirements.
- The remaining categories should be given third priority without explicit ranking, but should be closely monitored as needs evolve.

5. Recommendations

5.1 Recommendation One

Develop Performance-Based Standards for Digital Highway Measurement

One of the challenges encountered when developing this report was determining the current state-of-practice and technology readiness levels from the literature review and market survey. While there are some standards or requirements for the elements from the MET, there are no standards available for assessing how well technologies measure the MET elements.

It is well established that performance-based standards drive innovation. A case in point is the landmark NCHRP Report 350 (Ross Jr, Sicking, Zimmer, & Michie, 1993). This report was first issued in 1993 to update the crash testing of roadside hardware to address the increasing use of SUV's and light trucks among the driving public. Report 350 not only improved testing performance requirements for these heavier and higher center-of-gravity vehicles, but also created a tiered approach for safety requirements based upon traffic volume and speed. FHWA's mandated conformance to NCHRP Report 350 for national highways further drove innovation in roadside safety hardware (TRB, 2009).

Another illustrative example of the use of standards to drive innovation is the Department of Homeland Security's program to develop standards for Urban Search and Rescue (US&R) robots. This program, initiated in 2004, is charged with supporting the development of robotic technologies for US&R by creating a suite of performance-based standards relevant to US&R practitioners. The initial phase of the project involved a series of workshops that led to a robust requirements document for US&R robot technologies (Messina, et al., 2005). Researchers, manufacturers, and users then worked together to establish test protocols designed to demonstrate capabilities meeting these defined requirements. The effort is now part of the

ASTM E54 committee on Homeland Security Applications. The program has been effective in communicating user needs to industry, demonstrating and quantifying industry and research capabilities to end-users, and shaping near- and long-term development particularly targeting weaknesses exposed during performance tests.

As part of the SHRP-2 Highway Safety Research Projects, S-03-Roadway Measurement System Evaluation was charged with conducting "... an assessment of the state of the practice for mobile data collection of roadside and roadway characteristics and features related to safety analysis (TRB, 2008)." It is strongly recommended that FHWA extend this work to the development of open, consensus-based standards for Digital Highway Measurement. This effort, more than any other research investment, will help spur industry development and create a framework for state and local officials to specify and procure equipment and services.

5.2 Recommendation Two

Provide Standard Reference Data Sets for DHMS Algorithm Development

A critical part of the evaluation process is the availability of standard datasets against which different algorithms can be evaluated. The creation and maintenance of standard data sets for roadway inventory by a neutral party should be explored. The availability of standard reference data sets will allow for the evaluation of the robustness of the algorithms as it will assess the performance of the algorithms under varying conditions - conditions differing from those under which they were developed.

Standard Reference Data Sets should contain a large variety of data sets covering a broad range of experimental conditions (e.g., different light conditions for images acquired with cameras, various scanning densities of 3D imaging systems, different conditions of road signs and pavement markings, etc.). The Reference Data Sets should be grouped in two subsets: one exclusively for algorithm development and classifier training and one subset for testing only. In order to keep the training process separate from the testing process, the development/training subset should not contain data sets from the testing subset. Furthermore, both the training and testing subsets should be classified into positive examples (images containing the target) and negative examples (images without the target) as is required by contemporary machine learning algorithms.

Additionally, performance evaluation of different classification algorithms requires the development of objective and meaningful metrics. For any given algorithm, a combination of metrics will likely be required to provide a complete performance profile, for instance, the Receiver Operating Characteristics (ROC) graph shown in Figure 5.1. An ideal classifier is a step function with a True Positive Rate (TPR) equal to 1 for all False Positive Rate (FPR) > 0. In Figure 5.1, the two solid lines C1 and C2 correspond to two different classifiers: the C1 curve is above the C2 curve for all FPRs and therefore the classifier represented by C1 is better than the classifier represented by C2. It is a common mistake to assess an algorithm's performance by examining a single point on the ROC. For example, in Figure 5.1, the comparison of a single point from one curve (p1) to a single point from another curve (p2) is not the correct procedure. In fact, the TPR for p1 is less than the TPR for p2 and yet p1 belongs to the better classifier. Such a simple analysis considers only very narrow aspects of an algorithm's performance, and the results are unlikely to be duplicated in actual scenarios. For reference, the dashed diagonal line characterizes a classifier that randomly guesses the class (positive or negative).

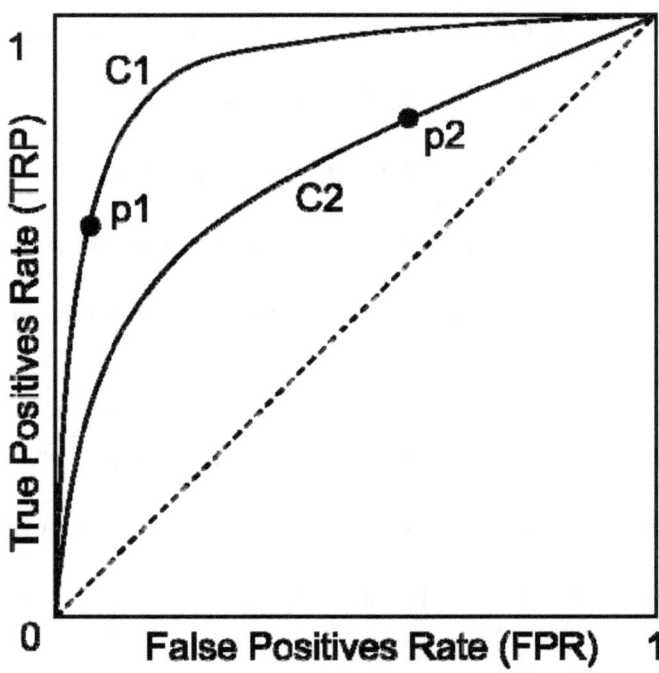

Figure 5.1. The Receiver Operating Curve (ROC) characterizes the performance of a classifier.

Maintaining Reference Data Sets together with properly defined metrics for classifier evaluation will encourage and accelerate the development of new recognition/classification algorithms. It will also ensure a solid basis for fair and objective comparison of COTS software.

Development and deployment of standard reference data sets provides opportunities for numerous entities (e.g., private companies, academia, research institutions) to simultaneously engage in algorithm development without the relatively high entry barrier of deploying data capture technologies. This strategy has proven effective in numerous science and technology initiatives to spur innovation and provide traceability for algorithm performance testing.

It is recommended that FHWA examine the high-priority MET elements, and lead the development of data sets which provide state-of-practice measurements of those elements (e.g., 2D and 3D camera data, laser-based 3D image data, etc.) with corresponding ground truth information. Ideally, this project would be an open-source endeavor where researchers worldwide could contribute new data sets to the project as new measurement technologies are developed.

5.3 Recommendation Three

Development of Object/Feature Recognition Algorithms

Since the early 1980s, extensive research has been conducted in the area of object and feature recognition, yet the development of a robust, general-purpose algorithm has remained elusive. Because of the quantity of information provided by cameras, the existence of such an algorithm could revolutionize highway asset management by allowing cameras to replace a host of commonly used sensor types. Most of the algorithms currently in use, however, are only effective in the restricted conditions under which they were developed. Extensions of these algorithms beyond their expected domains seems difficult. This observation agrees with the observation made for Item b in Table 1, i.e., the field of object and feature recognition is somewhat mature but automated extraction is not robust. At present, only face recognition is sufficiently robust for general implementation.

The original technique to object recognition, template matching, has shown promising performance for quite some time in the laboratory but several factors make this simple approach hard to translate to a deployment scenario. Due to minor differences between objects of the same type (known as "intra-class variation"), a new template would have to be trained for every potential object. Given the large number of objects in a roadway inventory, this is a formidable and uneconomical task. Viewing conditions also affect the quality of template matching: camera angle, viewing distance, lighting, and environmental factors all contribute to differences between the observed object and the equivalent object stored as a template.

Searching for a stored image of an object is not sufficiently robust for general use. Template matching algorithms are sensitive to variations in lighting, scale, color, and view angle, as well as the presence of occlusions. For this reason, the predominant contemporary approach to object recognition is to search for representative features of an object rather than searching for the object directly. The nature of these features vary between algorithms but is generally limited to low-level characteristic such as shape, color, or intensity gradients. Feature-based approaches have the benefit of being invariant to viewing distance, camera angle, and to some extent, occlusions. Typical features include SIFT (Scale Invariant Feature Transform), shape contexts, and Haar-like wavelets. These newer techniques still require object-specific training but their robustness to varying viewing conditions make them appealing for the application in question.

Future research in object and feature recognition should focus on adapting one or more of these newer techniques to the specific application. A representative training and testing set should be collected for each object type (see Recommendation Two). In practice, most effort in training classifiers is spent in collecting datasets and choosing low-level features (edges, colors, etc.) which best represent the target object.

Acknowledgements

Funding for this project was provided by FHWA Grant DTFH61-09-X-30011. The guidance and expertise provided by Mr. Lincoln Cobb of the FHWA Office of Safety R&D (HRDS-2) were invaluable to this study. The authors are also grateful to the other FHWA personnel who responded to the survey and provided feedback on this report. The support of the NIST Intelligent Systems Division in conducting the literature review / market analysis is greatly appreciated.

References

Blincoe, L., Seay, A., Zaloshnja, E., Miller, T., Romano, E., Luchter, S., et al. (2002). *The Economic Impact of Motor Vehicle Crashes, 2000*. Washington, DC: National Highway Traffic Safety Administration, U.S. Department of Transportation.

Cobb, L (2009). Future Vision of the DHMS - personal communication.

Council, Forrest M., Harkey, David L., Carter, Daniel L., & White, Bryon (2007). *Model Minimum Inventory of Roadway Elements - MMIRE* (No. FHWA-HRT-07-046): Federal Highway Administration.

DOT (2000). *Primer: GASB 34*.

Federal Railroad Administration (2007). *U. S. DOT National Highway-Rail Crossing Inventory, Policy, Procedures and Instructions for States and Railroads*.

FHWA (1995). *Recording and Coding Guide for the Structure Inventory and Appraisal of the Nation's Bridges* (No. FHWA-PD-96-C01). Washington D.C.

FHWA (2009a). TFHRC - Intersection Safety Retrieved June 30, 2009, from http://www.tfhrc.gov/safety/intersect htm

FHWA (2009b). TFHRC - Safety Research Retrieved June 30, 2009, from http://www.tfhrc.gov/safety/index htm#Intersections

Heron, M.P., Hoyert, D.L., Murphy, B.S., Xu, J., Kochaneck, M.A., & Tejada-Vera, B. (2009) *Deaths: Final data for 2006* (No. DHHS Publication No. (PHS) 2009-1120): Centers for Disease Control and Prevention, U.S. Department of Health and Human Services.

Messina, E., Jacoff, A., Scholtz, J., Schlenoff, C., Huang, H.-M., Lytle, A., and Blitch, J. (2005), Statement of Requirements for Urban Search and Rescue Robot Performance Standards, Preliminary Report, May 13, 2005, National Institute of Standards and Technology, Gaithersburg, MD.

NCDOT. Inventory Asset Data Collection Elements. *National Conference: Highway Asset Inventory and Data*

Collection, from http://www.itre ncsu.edu/ncassetmgmtconf/downloads/RoadsideDataElementTable.pdf

NCDOT. NCDOT Pavement Condition Survey. *National Conference: Highway Asset Inventory & Data Collection*, 2009, from http://www.itre.ncsu.edu/ncassetmgmtconf/downloads/PavementsDataElementTable.pdf

NHTSA (2000). *Traffic Safety Facts 2000*. Washington, DC: National Highway Traffic Safety Administration, U.S. Department of Transportation.

NHTSA (2009). *Traffic Safety Facts 2007*. Washington, DC: National Highway Traffic Safety Administration, U.S. Department of Transportation.

Ross Jr, H.E., Sicking, D.L., Zimmer, R.A., & Michie, J.D. (1993). *NCHRP Report 350: Recommended Procedures for the Safety Performance Evaluation of Highway Features*.

Rumar, K. (1985). The role of perceptual and cognitive filters in observed behavior. *Human behavior and traffic safety*, 151-165.

TRB (2008). *SHRP-2 Safety Program Brief*: Transportation Research Board, National Academy of Sciences.

TRB (2009). *Crash standards save lives, spur innovation*: Transportation Research Board, National Academy of Sciences.

Trentacoste, M.F. (2006). Digital Highway Measurement System (Presentation to the 5th International Visualization In Transportation Symposium). McLean, VA: Office of Safety Research and Development, Turner Fairbank Highway Research Center, FHWA.

Vandervalk, Anita (2008, July 17, 2008). *SHRP 2 Project S-03, Task 2, Determination & Prioritization of Data Elements*. Paper presented at the Fourth SHRP Safety Symposium.

Appendix A: Digital Highway Measurement Technology Database

A Digital Highway Measurement Technology Database was developed during this study. The two primary objectives of this effort were: 1) to enable FHWA and their research partners to interactively analyze the data from both the Master Element Table (MET) and Technology Table (TT), and 2) to create a collaborative environment for continuous maintenance and updates. The main functionalities of the database are:

- Maintain records
    - Collaboratively add, delete, and edit records

- Browse the element and technology records

- Identify technologies that address a particular element

- Identify elements that are addressed by a particular technology

- Sort by element and/or technology attribute(s):
    - An example of an attribute is technology readiness level (e.g., COTS, Prototype, R&D, Concept).

- Filter by element and/or technology attribute(s):
    - For example, within an attribute, filter to show only the technologies for a given technology readiness level (e.g., COTS only).

- Generate formatted reports (see sample outputs below) for each element and technology

- Provide a mechanism to generate custom reports

Sample information from the database for one element (Element Number 22) and one Technology (T0001) follows.

Element Number: 22
Element Name: Surface Friction

Source of Element:
MMIRE (Council, Harkey, Carter, & White, 2007)

Level 1 Descriptor:
I. ROADWAY SEGMENT

Level 2 Descriptor:
I.c. Segment Cross Section

Level 3 Descriptor:
I.c.1. Surface Descriptors

Description:
None given in MMIRE.

Surface friction has two major components (Noyce, Bahia, Yambó, & Kim, 2005):
1. Microtexture refers to irregularities in the surfaces of the stone particles (fine-scale texture) that affect adhesion (molecular bond shearing).
2. Macrotexture refers to the larger irregularities in the road surface (coarse-scale texture) that affects hysteresis (energy dissipation from tire deformation)

Additional secondary components (Noyce, et al., 2005):
1. Megatexture describes irregularities that can result from rutting, potholes, patching, surface stone loss, and major joints and cracks
2. Roughness refers to surface irregularities larger than megatexture that also affects rolling resistance, in addition to ride quality and vehicle operating costs.

There are three predominant measurement techniques (WSDOT, 2009 Module 9: Pavement Evaluation, Section 4: Skid Resistance):
1. Locked wheel tester (AASHTO T 242 or ASTM E-274)
   Consists of a locked wheel skidding along a wet surface to measure friction resistance. This is the most common method.
2. Spin up tester
   With a similar setup as the locked wheel tester, an unlocked wheel is lowered onto the pavement and the spin up time is measured. This requires no force sensors and reduces wear on the test wheel.
3. Surface texture measurement
   Surface texture is measured (see Element 23) and a mathematical model correlates surface texture to surface friction.

What to record:
Surface friction indicator
- "Measured skid number on the segment or general indication of wet-surface friction (e.g., high, medium, low)" (Council, et al., 2007)

Lock Wheeled Skid Tester *(WSDOT, 2009 Module 9: Pavement Evaluation, Section 4: Skid Resistance)*

Out of Scope?
   No

Derived from Other Elements?
   No

Measured from a Moving Vehicle?
   Yes

Similar to Other Element(s)?
   No

Ease of Data Collection (from MMIRE):
   Difficult

Levels of Uncertainty Required:
   ASTM E274-06 states that acceptable precision of Skid Number (SN) units can be stated in the form of repeatability. Acceptable standard deviation of 2 SN units was obtained from ASTM tests with varying instruments at different speeds (ASTM, 2006b).

   ASTM E274 (ASTM, 2006b), E1337 (ASTM, 2008), and E1859 (ASTM, 2006a) require the following specifications for the force-measuring transducer for a locked wheel test setup. The transducer shall measure the tire-road interface force with minimal inertial effects. It shall provide an output directly proportional to the force with hysteresis less than 1 % of the applied load, nonlinearity less than 1 % of the applied load up to the max expected loading, and sensitivity to any expected cross-axis loading or torque loading less than 1 % of the applied load.

   ASTM E274 (ASTM, 2006b) requires the torque-measuring transducer in a locked wheel setup to provide an output directly proportional to the torque with hysteresis less than 1 % of the applied load, nonlinearity less than 1 % of the applied load up to the max expected loading, and sensitivity to any expected cross-axis loading or torque loading less that 1 % of the applied load.

ASTM E274 (ASTM, 2006b) and E1337 (ASTM, 2008) require the vehicle speed-measuring transducers to have a speed resolution accuracy of +/- 1.5 % of the indicated speed or +/- 0.8km/h, whichever is greater.

ASTM E274 (ASTM, 2006b) requires an overall system accuracy of +/- 1.5 % of the applied load from 900 N to full scale. ASTM E1337 (ASTM, 2008) requires an overall system accuracy of +/- 1.5 % of the applied load from 890 N to full scale.

Applicable Measurement Technologies:
T002, T008, T010, T013, T019, T028, T056

Potential Measurement Technologies:
T009 – if method to determine surface friction based on macro- and microtexture parameters is implemented.

MMIRE Priority:
1

MMUCC Element?
No

Intersection Element?
No

Roadway Alignment Element?
No

References:

ASTM (2006a). Standard Test Method for Friction Coefficient Measurements Between Tire and Pavement Using a Variable Slip Technique, *E1859-97*.

ASTM (2006b). Standard Test Method for Skid Resistance of Paved Surfaces Using a Full-Scale Tire, *E274-06*.

ASTM (2008). Standard Test Method for Determining Longitudinal Peak Braking Coefficient of Paved Surfaces Using Standard Reference Test Tire, *E1337-90*.

Council, Forrest M., Harkey, David L., Carter, Daniel L., & White, Bryon (2007). *Model Minimum Inventory of Roadway Elements - MMIRE* (No. FHWA-HRT-07-046): Federal Highway Administration.

Noyce, DA, Bahia, HU, Yambó, JM, & Kim, G (2005). Incorporating road safety into pavement management: Maximizing asphalt pavement surface friction for road safety improvements. *Midwest Regional University Transportation Center Traffic Operations and Safety (TOPS) Laboratory, Draft Literature Review and State Surveys (April 2005)*.

WSDOT (2009). Washington State Department of Transportation Pavement Guide, Module 9: Pavement Evaluation, Section 4: Skid Resistance Retrieved 05/08/2009, 2009, from http://training.ce.washington.edu/wsdot/Modules/09_pavement_evaluation/09-4_body.htm

Technology Number: T0001
Technology Proprietor Name: Fugro-Roadware
Product Name: ARAN

General Technology Category(s):
    Terrestrial System
    GPS
    Inertial System
    Camera-based 2D Imaging
    Laser-based 2D Imaging
    Other

Functional Classification(s):
    Location measurement
    Object/feature recognition
    Spatial measurement

Information:
    Vehicle measures longitudinal and transverse road profiles (IRI, ride number, rut), velocity and distance traveled (DMI). Profiler conforms to: AASHTO PP 38-00, AASHTO PP 49-03, AASHTO PP 50-03, AASHTO MP 11, ASTM E950-98, ASTM E1705/E1705M

    Vehicle (Wisecrack) has automated crack detection - reports crack type, severity and location - compliant with AASHTO and LTPP (SHRP) Protocols.

    Smart Geometrics is a vehicle-mounted subsystem that utilizes a patented control algorithm and a combination of gyroscopes and software to measure the crossfall, transverse profile, vertical alignment (grade) and horizontal alignment (curve radius) of the roadway.

    Vehicle (Surveyor 2) asset surveying - determine the linear position, measurements, X Y and Z location and other user-defined attributes of roadside assets from geo-referenced digital images.
    Source: (Roadware, 2009)

Achievable/Potential Level of Measurement Uncertainty:
    Cracks >= 1 mm, Rut depth accuracy of 1.5 mm, GPS of 1 m or 5 m

Technology Readiness:
    Commercial-off-the-shelf

Addresses Elements:
    5, 7, 23, 26, 32, 33, 35, 37, 42, 47, 59, 68, 70, 138, 149, 177, 197, 198, 204, 230, 299, 302, 303, 304, 305, 311

Elements that Could be Addressed:
    28 - The system can report crack type and severity, IRI, and rut depth - depending on the definition of pavement condition, the system can be used to evaluate this element.

34 - Polished aggregates may be detected (manually or automatically if an algorithm were developed) from digital images.

36 - Water bleeding may be detected (manually or automatically if an algorithm were developed) from digital images.

39 - uncertain if the results of oxidation is detectable from digital images or not, or if mechanical properties of the pavement need to be measured

40 - Bleeding may be detected (manually or automatically if an algorithm were developed) from the digital images.

41 - Pumping may be detected (manually or automatically if an algorithm were developed) from the digital images.

53 - Blowups can be detected from digital images (pavement module).

54 - Element can be detected from digital images and the length can be measured using the profiling system or from a calibrated image. Will need good resolution to measure separation to the millimeter level.

55 - Element can be detected from digital images and distances/lengths can be obtained from calibrated images or profiling system.

143, 144 - 2D images may be needed for initial inlet identification, 3D LADAR for more accurate; difficult task for software recognition due to expected large variation in shape, geometry, etc. of different inlets

149 - Based on 2008 Highway Asset Inventory and Data Collection NCDOT wkshop, this element was obtained with this technology.

150 - Driveways and driveway type may be detected from digital images

161 - Barrier condition may be assessed from video. Recognition algorithm will need to be developed.

174, 175, 176 - obstacles can be identified from video and offset from road can be measured - (Surveyor 2 vehicle), and the location can be obtained from GPS

178 - element can be identified from digital images which is tied to GPS

194 - curb/sidewalk width may be obtained from their Surveyor 2 module

300 - curve length could be determined with DMI & gyroscopes;

313, 314, 318 - can be derived from the raw sensor data, given the appropriate algorithms

Data Processing Speed:
Offline

Data Processing Level of Automation:
Unknown

Contact Information for Technology:
Fugro Roadware Inc. (USA)
3104 Northside Avenue
Richmond, Virginia 23228
USA
Toll Free: +1 800 828-ARAN (2726)

Other information:
1. 2003 report (Javidi, et al., 2003) on the performance of WiseCrack and proposed improvements.

2. Roadware was one of the vehicles evaluated in (Mullis, 2005).

3. 2008 Highway Asset Inventory & Data Collection National Workshop ("2008 Highway Asset Inventory & Data Collection, Roadside Manual and Vendor Data Sets," Roadside Manual) - Roadware was one of the vehicles that collected data.

4. 2003 report (Sokolic, 2003) - Early version of ARAN was evaluated.

References:

2008 Highway Asset Inventory & Data Collection, Roadside Manual and Vendor Data Sets Retrieved June 25, 2009, from http://www.itre.ncsu.edu/NCassetMgmtConf/vendordatasets.html
Javidi, B., Stephens, J., Kishk, S., Naughton, T., McDonald, J., & Isaac, A. (2003). *Pilot for Automated Detection and Classification of Road Surface Degradation Features* (No. JHR 03-295): Univ. of Connecticut.
Mullis, C., Reid, J. Brooks, E., Shippen, N. (2005). *Automated Data Collection Equipment for Monitoring Highway Condition* (No. SPR 332 and FHWA-OR-RD-05-10): Oregon DOT and FHWA.
Roadware (2009). Fugro Roadware - ARAN, from http://www.roadware.com/products_services/aran/
Sokolic, Ivan (2003). *Criteria to Evaluate the Quality of Pavement Camera Systems in Automated Evaluation Vehicles.* Unpublished Masters Thesis.

Appendix B: Technology Search Procedure

Assumptions:
- Applicable technologies and relevant research is assumed to be published in some form on the internet and reachable through a Google search.

Instructions:
- Search time is generally limited to two hours per element.

- Use google.com as the search engine and set the following Google search preferences:
    - Language: English
    - Safe Search Filtering: Moderate
    - Results displayed per page: 10

- Use the file "Elements_to_search.xls" to determine which elements you have been assigned.

- Use the file template "element_technology_search.xlt" to document your search
    - Save the file with the element number as the filename (e.g., if you're working on element number 171, the filename would be "171.xls")

- Do not include papers that are published before 2004 as they are likely out-dated due to the rapid advance in technology (this is especially true in the image processing field) - unless there is a reason to.

- Record the following with each new search term in the above Excel file:
    - The exact search terms used for each search
    - The number of results for each search

- The information recorded for each new technology is shown on the last page of these instructions.

Procedure:

1. Look at the "Technology Table" and try to find the element # in column S. If the element exists, then go to Step 4 (quick search for other technologies).

2. If the element is not listed, identify the technology classification(s) required for the element under investigation and search "Technology Table" on Google Docs (columns X through AD). If a technology already exists in the database that meets the requirements of the element, then add your element number to it and go to Step 5 (moderate search for other technologies).

3. If Steps 1 and 2 do not turn up any applicable technologies, then do a thorough search (six searches for each search term). This consists of:

    a. Google search on the search term[11] in quotes (i.e., "search term") plus the words "road" then "highway" with the qualifiers identified in the spreadsheet (e.g., (detect OR measure)) with each search and the acronym "DOT" for the searches that do not include phrases (i.e., words in quotation marks). The six searches become more general as you progress.

    E.g., if the element 171 is *Sign Type*, then you would type the following six searches in Google exactly as they appear (without the roman numerals of course):
    i. "sign type" road (detect OR recognize)
    ii. "sign type" highway (detect OR recognize)
    iii. sign type road (detect OR recognize) DOT
    iv. sign type highway (detect OR recognize) DOT
    v. sign type (detect OR recognize) DOT
    vi. sign (detect OR recognize) DOT

    b. For each search, look through the first three pages of results. If no applicable technologies are found, look through pages 4 and 5.
    c. Use scholar.google.com and go through three pages of results using the most appropriate term based on your own judgment and record the term used.

4. For a quick search, use two search terms (one general and one specific if possible) and follow Steps 3b and 3c.

5. For a moderate search, use four search terms (two general and two specific if possible) and follow Steps 3b and 3c.

---

[11] See Table 1. Also, the search term in many cases is equivalent to the element name.

Information that should be recorded for each technology identified

- **General Technology Category:**
    - *Terrestrial system*
    - *Airborne system*
    - *Laser-based 3D imaging*
    - *Camera-based 3D imaging*
    - *Global positioning system*
    - *Inertial system*
    - *Software only*
    - *Camera-based 2D digital imaging*
    - *Laser-based 2D imaging*
    - *Other*
- **Technology proprietor name:** *(Company/researcher name)*
- **Product name (if applicable):** *(E.g., Bentley's "InRoads" software)*
- **Description:** *(Brief description of what the technology is)*
- **Achievable (or potential) level of measurement uncertainty:**
- **Technology readiness:**
    - *Concept*
    - *R&D*
    - *Prototype/Demonstration*
    - *Commercial-off-the-shelf*
- **Element #:**
- **Data processing speed:** *(Offline/Real-time/Unknown)*
- **Data processing level of automation:** *(Manual/Automatic/Semi-automatic/Unknown)*
- **Contact information for technology:** *(Company contact information)*
- **Other information:**
- **Functional classification:**
    - *Location measurement*
    - *Object/feature recognition*
    - *Sign-type recognition*
    - *Character recognition (words and numbers on signs, etc.)*
    - *Spatial measurement*
    - *Material optical property measurement (color, retro-reflectivity, etc.)*
    - *Material property recognition (concrete, asphalt, wood, metal, paint, thermoplastic, etc.)*

Appendix C: Element Priority Survey

The elements in Table C.1 were identified based on the criteria discussed in Section 4.3 and were included in the survey sent to FHWA. The survey results are summarized in Table C.2. The numbers in columns A and B are the numbers of responses for each choice. In the comments column (C) in Table C.2, the numbers at the beginning of the comments correspond to the respondent numbers.

Table C.1. High Priority Elements in Survey

	Data Elements and Categories	A. This element has a high safety priority.				B. Technologies exist to measure this element.			C. Comments
		Agree	Disagree	Strongly Disagree	Don't Know	Yes	Possible	No	
I. Pavement Condition Descriptors									
1	Polished Aggregate - [Record square feet of affected surface area]	Agree	Disagree	Strongly Disagree	Don't Know	Yes	Possible	No	
2	Water Bleeding - [Record the number of occurrences of water bleeding and the length in meters of affected pavement with a minimum length of 3.28 feet]	Agree	Disagree	Strongly Disagree	Don't Know	Yes	Possible	No	
3	Oxidation - [Only plant mix surfaces are rated for oxidation. Determine presence of oxidation]	Agree	Disagree	Strongly Disagree	Don't Know	Yes	Possible	No	
4	Bleeding - [Record square feet of surface area at each severity level]	Agree	Disagree	Strongly Disagree	Don't Know	Yes	Possible	No	
5	Pumping - [Record the number of occurrences of pumping and the length in feet of affected pavement]	Agree	Disagree	Strongly Disagree	Don't Know	Yes	Possible	No	
6	Shoving - [Record number of occurrences and square feet of affected surface area]	Agree	Disagree	Strongly Disagree	Don't Know	Yes	Possible	No	
7	Scaling - [Record the number of occurrences and the square meters of affected area]	Agree	Disagree	Strongly Disagree	Don't Know	Yes	Possible	No	

	Data Elements and Categories	A. This element has a high safety priority.				B. Technologies exist to measure this element.		C. Comments
8	Lane-to-Shoulder Separation - [Record to the nearest millimeter at intervals of 50 feet along the lane-to-shoulder joint. Indicate whether the joint is well-sealed (yes or no) at each location]	Agree	Disagree	Strongly Disagree	Don't Know	Yes	Possible	No
9	Transverse Construction Joint Deterioration - [Record number of construction joints at each severity level]	Agree	Disagree	Strongly Disagree	Don't Know	Yes	Possible	No
10	Lane Add Point - [GPS or Reference Post of Start of a FULL lane width]	Agree	Disagree	Strongly Disagree	Don't Know	Yes	Possible	No
11	Exclusive Left Turn Lane Length - [Exclusive left turn lane length]	Agree	Disagree	Strongly Disagree	Don't Know	Yes	Possible	No
12	Auxiliary Lane Presence/Type - [Presence or type of auxiliary lane]	Agree	Disagree	Strongly Disagree	Don't Know	Yes	Possible	No
13	Auxiliary Lane Length - [Length of auxiliary lane]	Agree	Disagree	Strongly Disagree	Don't Know	Yes	Possible	No
14	HOV Lanes - [Presence of HOV lanes in segment]	Agree	Disagree	Strongly Disagree	Don't Know	Yes	Possible	No
15	HOV Lane Types - [HOV lane types]	Agree	Disagree	Strongly Disagree	Don't Know	Yes	Possible	No
16	Reversible Lanes - [Number of reversible lanes present on segment]	Agree	Disagree	Strongly Disagree	Don't Know	Yes	Possible	No
17	Presence/Type of Bicycle Facility - [Presence or type of bicycle facility on segment]	Agree	Disagree	Strongly Disagree	Don't Know	Yes	Possible	No
18	Number of Peak Hour Lanes - [Number of through lanes used in peak period in the peak direction]	Agree	Disagree	Strongly Disagree	Don't Know	Yes	Possible	No
II. Shoulder, Median, and Roadside Descriptors								
19	Right Shoulder Type - [Shoulder type on right side of road in direction of inventory]	Agree	Disagree	Strongly Disagree	Don't Know	Yes	Possible	No
20	Right Paved Shoulder Width - [Width of paved portion of right shoulder]	Agree	Disagree	Strongly Disagree	Don't Know	Yes	Possible	No
21	Shoulder Rumble Strip Presence - [Presence of shoulder rumble strip]	Agree	Disagree	Strongly Disagree	Don't Know	Yes	Possible	No
22	Rumble Strip Type - [Rumble strip type if present]	Agree	Disagree	Strongly Disagree	Don't Know	Yes	Possible	No
23	Location of Rumble Strip - Begin - [GPS or reference post of start]	Agree	Disagree	Strongly Disagree	Don't Know	Yes	Possible	No
24	Rumble Strip Offset - [From edge of lane]	Agree	Disagree	Strongly Disagree	Don't Know	Yes	Possible	No

	Data Elements and Categories	A. This element has a high safety priority.				B. Technologies exist to measure this element.		C. Comments
25	Sidewalk is separated from edge of road - [Yes/No]	Agree	Disagree	Strongly Disagree	Don't Know	Yes	Possible	No	
26	Curb Type - [Curb type]	Agree	Disagree	Strongly Disagree	Don't Know	Yes	Possible	No	
27	Curb Location - [GPS coordinates of beginning and end point]	Agree	Disagree	Strongly Disagree	Don't Know	Yes	Possible	No	
28	Curb Blockage	Agree	Disagree	Strongly Disagree	Don't Know	Yes	Possible	No	
29	Curb Damage	Agree	Disagree	Strongly Disagree	Don't Know	Yes	Possible	No	
30	Median Opening Location - [GPS of first corner in direction of travel]	Agree	Disagree	Strongly Disagree	Don't Know	Yes	Possible	No	
31	Median Barrier Type - [Soil, paved (striped), paved (barrier), raised curb, None, Other]	Agree	Disagree	Strongly Disagree	Don't Know	Yes	Possible	No	
32	Barrier beginning point location - [GPS or Reference Post of Beginning of entire barrier system]	Agree	Disagree	Strongly Disagree	Don't Know	Yes	Possible	No	
33	Barrier offset - beginning - [From edge of lane]	Agree	Disagree	Strongly Disagree	Don't Know	Yes	Possible	No	
34	Barrier height - [Height from ground surface in inches]	Agree	Disagree	Strongly Disagree	Don't Know	Yes	Possible	No	
35	Barrier post type - [Strong Post (Metal), Weak Post (Metal), Wooden Post, N/A, Other]	Agree	Disagree	Strongly Disagree	Don't Know	Yes	Possible	No	
36	Barrier offset bracket - [Yes/No]	Agree	Disagree	Strongly Disagree	Don't Know	Yes	Possible	No	
37	Barrier rub rail - [Yes/No]	Agree	Disagree	Strongly Disagree	Don't Know	Yes	Possible	No	
38	Barrier end treatment type - beginning - [Impact Attenuator, Buried End, Terminal End, Fist, Bridge Connection, None, Other]	Agree	Disagree	Strongly Disagree	Don't Know	Yes	Possible	No	
39	Median Left Turn Lane Type - [Type of left turn lane in median]	Agree	Disagree	Strongly Disagree	Don't Know	Yes	Possible	No	
40	Median Left Turn Lane Width - [Width of median left turn lane]	Agree	Disagree	Strongly Disagree	Don't Know	Yes	Possible	No	
41	Drop Inlet Location - [GPS coordinates]	Agree	Disagree	Strongly Disagree	Don't Know	Yes	Possible	No	
42	Drop Inlet Blockage	Agree	Disagree	Strongly Disagree	Don't Know	Yes	Possible	No	
43	Roadside Clearzone Width - [Roadside clearzone width]	Agree	Disagree	Strongly Disagree	Don't Know	Yes	Possible	No	

	Data Elements and Categories	A. This element has a high safety priority.				B. Technologies exist to measure this element.		C. Comments
44	Roadside Clearzone Cross Slope (Assume Sideslope = Clearzone Cross Slope) - [Sideslope]	Agree	Disagree	Strongly Disagree	Don't Know	Yes	Possible No	
45	Driveway Type (MMIRE = Driveway Information) - [Residential, Farm, Retail/Commercial, Industrial]	Agree	Disagree	Strongly Disagree	Don't Know	Yes	Possible No	
46	Barrier Condition - [Functional or not]	Agree	Disagree	Strongly Disagree	Don't Know	Yes	Possible No	
47	Attenuator present (barrier beginning) - [Yes/No]	Agree	Disagree	Strongly Disagree	Don't Know	Yes	Possible No	
48	Attenuator type -	Agree	Disagree	Strongly Disagree	Don't Know	Yes	Possible No	
49	Attenuator Condition - [Functional or not]	Agree	Disagree	Strongly Disagree	Don't Know	Yes	Possible No	
50	Offset of roadside obstacle - [From edge of lane]	Agree	Disagree	Strongly Disagree	Don't Know	Yes	Possible No	
51	Location of roadside obstacle - [GPS or reference post of each obstacle]	Agree	Disagree	Strongly Disagree	Don't Know	Yes	Possible No	
III. Bridge Descriptors								
52	Bridge Begin Location (MMIRE: Bridge Descriptors for Bridges in Segment) - [GPS coordinates]	Agree	Disagree	Strongly Disagree	Don't Know	Yes	Possible No	
53	Bridge approach Slab Settlement Exists - [Yes/No]	Agree	Disagree	Strongly Disagree	Don't Know	Yes	Possible No	
54	Offset of bridge rail - [Distance from edge of lane]	Agree	Disagree	Strongly Disagree	Don't Know	Yes	Possible No	
55	Bridge Median - [Whether the median is non-existent, open or closed. The median is closed when the area between the 2 roadways at the structure is bridged over and is capable of supporting traffic. All bridges that carry either 1-way traffic or 2-way traffic separated only by a centerline have no median]	Agree	Disagree	Strongly Disagree	Don't Know	Yes	Possible No	
56	Bridge transitions - [The transition from approach guardrail to bridge railing requires that the approach guardrail be firmly attached to the bridge railing. It also requires that the approach guardrail be gradually stiffened as it comes closer to the bridge railing. The ends of curbs and safety walks need to be gradually tapered out or shielded]	Agree	Disagree	Strongly Disagree	Don't Know	Yes	Possible No	

Data Elements and Categories	A. This element has a high safety priority.			B. Technologies exist to measure this element.		C. Comments
	Agree	Disagree	Strongly Disagree	Don't Know	Yes	Possible	No	
57	Type of Service on Bridge - [Classification: Highway, Railroad, Pedestrian-bicycle, Highway-railroad, Highway-pedestrian, Overpass structure at an interchange or second level of a multilevel interchange, Third level (Interchange), Fourth level (Interchange), Building or plaza, Other]	Agree	Disagree	Strongly Disagree	Don't Know	Yes	Possible	No	
58	Left Curb/Sidewalk Width - [Widths of the left and right curbs or sidewalks to nearest tenth of a meter. "Left" and "Right" should be determined on the basis of direction of the inventory]	Agree	Disagree	Strongly Disagree	Don't Know	Yes	Possible	No	
IV. Railroad Crossing Descriptors							
59	Number of Tracks - [Number of railroad tracks]	Agree	Disagree	Strongly Disagree	Don't Know	Yes	Possible	No	
60	Railroad Crossing Control Type - [Crossbucks, gates, flashing lights, signal]	Agree	Disagree	Strongly Disagree	Don't Know	Yes	Possible	No	
61	Railroad crossing number - [U.S. DOT inventory crossing number]	Agree	Disagree	Strongly Disagree	Don't Know	Yes	Possible	No	
62	Railroad crossing Four-quadrant (or Full Barrier) Gates - [Whether or not four-quadrant (or full barrier) gates are present at the crossing. Full barrier gates apply in the case of 1-way streets or where the gate arms reach across the entire roadway]	Agree	Disagree	Strongly Disagree	Don't Know	Yes	Possible	No	
63	Railroad crossing Cantilevered (or Bridged) Flashing Lights - [The number of cantilevered (or bridged) flashing lights in the appropriate block. Separate cantilevered flashers from those over traffic lanes and those not reaching the roadway (over only parking lanes, turnout lanes, or shoulders). Count individual cantilever units; do not count the flasher head pairs mounted on the units]	Agree	Disagree	Strongly Disagree	Don't Know	Yes	Possible	No	
64	Railroad crossing Wigwags - [The number of wigwag signals]	Agree	Disagree	Strongly Disagree	Don't Know	Yes	Possible	No	
65	Railroad crossing Bells - [The number of all bells, if present, that are either alone or in conjunction with other train activated warning devices]	Agree	Disagree	Strongly Disagree	Don't Know	Yes	Possible	No	

	Data Elements and Categories	A. This element has a high safety priority.				B. Technologies exist to measure this element.			C. Comments
		Agree	Disagree	Strongly Disagree	Don't Know	Yes	Possible	No	
66	Crossing Surface (on Main Line) - [Type of crossing surface (if there are multiple tracks which have different types of surfaces, indicate the lower grade surface material): Timber, Sectional Treated Timber, Full Wood Plank, Asphalt, Asphalt and Flange, Concrete, Concrete Slab, Concrete and Rubber, Rubber, Metal, Metal Sections, Other Metal, Unconsolidated, Other]	Agree	Disagree	Strongly Disagree	Don't Know	Yes	Possible	No	
V. Traffic Operations/Control Descriptors									
67	One/Two-Way Operations - [Whether the segment operates as a one- or two-way roadway]	Agree	Disagree	Strongly Disagree	Don't Know	Yes	Possible	No	
68	On-Street Parking Type - [On-street parking type]	Agree	Disagree	Strongly Disagree	Don't Know	Yes	Possible	No	
69	Beginning of on-street parking - [GPS or Reference of start]	Agree	Disagree	Strongly Disagree	Don't Know	Yes	Possible	No	
70	Roadway Lighting Presence - [Yes/No]	Agree	Disagree	Strongly Disagree	Don't Know	Yes	Possible	No	
71	Location of street lighting - [GPS or reference post of pole]	Agree	Disagree	Strongly Disagree	Don't Know	Yes	Possible	No	
72	Toll Facility? - [Toll facility indicator]	Agree	Disagree	Strongly Disagree	Don't Know	Yes	Possible	No	
73	Edgeline Marking Width - [To the nearest inch]	Agree	Disagree	Strongly Disagree	Don't Know	Yes	Possible	No	
74	Edgeline Marking Color -	Agree	Disagree	Strongly Disagree	Don't Know	Yes	Possible	No	
75	Edgeline Location of marking - begin - [GPS or Reference of start]	Agree	Disagree	Strongly Disagree	Don't Know	Yes	Possible	No	
76	Edgeline Marking Material Type - [Paint, Thermoplastic, Polyurethane]	Agree	Disagree	Strongly Disagree	Don't Know	Yes	Possible	No	
77	Edgeline Marking offset - [Offset of each type of line (center, lane & edge) from right edge of pavement]	Agree	Disagree	Strongly Disagree	Don't Know	Yes	Possible	No	
78	Edgeline marking type	Agree	Disagree	Strongly Disagree	Don't Know	Yes	Possible	No	
VI. Horizontal and Vertical Curve Descriptors									
79	Curve Feature Type - [Type of horizontal alignment feature being described in the data record]	Agree	Disagree	Strongly Disagree	Don't Know	Yes	Possible	No	
80	Sight Distance (Stopping) - [Report sight distance at 0.01-mile intervals for Horizontal Curves from PC to PT]	Agree	Disagree	Strongly Disagree	Don't Know	Yes	Possible	No	

	Data Elements and Categories	A. This element has a high safety priority.				B. Technologies exist to measure this element.		C. Comments
		Agree	Disagree	Strongly Disagree	Don't Know	Yes	Possible No	
81	Sight Distance (Stopping) - [Report sight distance at 0.01-mile intervals for Vertical Curves from PC to PT]	Agree	Disagree	Strongly Disagree	Don't Know	Yes	Possible No	
VII. Intersection and Interchange Descriptors								
82	Type of Intersection/Junction - [Type of junction being described in the data record]	Agree	Disagree	Strongly Disagree	Don't Know	Yes	Possible No	
83	Location Identifier for Road 1 Crossing Point - [Location on the first intersecting route (e.g., route-milepost)]	Agree	Disagree	Strongly Disagree	Don't Know	Yes	Possible No	
84	Intersection/Junction No. of Legs - [Intersection/junction no. of legs]	Agree	Disagree	Strongly Disagree	Don't Know	Yes	Possible No	
85	Intersection/Junction Geometry - [Intersection/junction geometry]	Agree	Disagree	Strongly Disagree	Don't Know	Yes	Possible No	
86	Intersection Skew Angle - [Angle from perpendicular of intersection of the roads]	Agree	Disagree	Strongly Disagree	Don't Know	Yes	Possible No	
87	Intersection/Junction Offset - [Whether crossroad approach centerlines are directly opposed or offset by some distance]	Agree	Disagree	Strongly Disagree	Don't Know	Yes	Possible No	
88	Intersection/Junction Offset Distance - [Distance that approach centerlines are offset]	Agree	Disagree	Strongly Disagree	Don't Know	Yes	Possible No	
89	Type of signalized intersection - [Standard, Protected Turn, Permitted Turn]	Agree	Disagree	Strongly Disagree	Don't Know	Yes	Possible No	
90	Flashing beacon present [at stop-controlled intersection] - [Yes/No (Flashing yellow/red beacon)]	Agree	Disagree	Strongly Disagree	Don't Know	Yes	Possible No	
91	Number of Intersection/Junction Quadrants With Limited Sight Distance - [Number of intersection/junction quadrants with limited sight distance]	Agree	Disagree	Strongly Disagree	Don't Know	Yes	Possible No	
92	Roundabout-Inscribed Diameter - [Distance between the outer edges of the circulatory roadway]	Agree	Disagree	Strongly Disagree	Don't Know	Yes	Possible No	
93	Channelization exists on approach - [Yes/No]	Agree	Disagree	Strongly Disagree	Don't Know	Yes	Possible No	
94	Pedestrian Signalization Type - [Type of pedestrian signalization on approach]	Agree	Disagree	Strongly Disagree	Don't Know	Yes	Possible No	
95	Roundabout-Entry Radius - [Minimum radius of curvature of the curb on the right side of the entry]	Agree	Disagree	Strongly Disagree	Don't Know	Yes	Possible No	
96	Roundabout-Pedestrian Facility - [Type of pedestrian crossing facility on this approach to roundabout]	Agree	Disagree	Strongly Disagree	Don't Know	Yes	Possible No	

	Data Elements and Categories	A. This element has a high safety priority.				B. Technologies exist to measure this element.		C. Comments
		Agree	Disagree	Strongly Disagree	Don't Know	Yes	Possible	No
97	Roundabout-Splitter Island Width - [Width of the splitter island separating entry and exit legs (measured at the inscribed circle)]	Agree	Disagree	Strongly Disagree	Don't Know	Yes	Possible	No
98	Ramp Location - [GPS Coordinates or Reference Post of gore area]	Agree	Disagree	Strongly Disagree	Don't Know	Yes	Possible	No
99	Ramp Length - [Length of ramp]	Agree	Disagree	Strongly Disagree	Don't Know	Yes	Possible	No
100	Location Identifier For Roadway at Beginning Ramp Terminal - [Location on the roadway at the beginning ramp terminal (e.g., route-milepost for that roadway) if the ramp connects with a roadway at that point]	Agree	Disagree	Strongly Disagree	Don't Know	Yes	Possible	No

Table C.2. Survey Results

	Data Elements and Categories	A. This element has a high safety priority.				B. Technologies exist to measure this element.			C. Comments
		\multicolumn{7}{l	}{I. Pavement Condition Descriptors}		
1	Polished Aggregate - [Record square feet of affected surface area]	5	0	0	5	2	3	3	1. It is possible that a measurement of overall reflectance from a line scanning laser can provide a reasonable estimate.; 8. Skid trailers; 10. There are several vendors trying to promote their digital imagery technology as capable of collecting the roadway elements identified. The difficulty is automating and integrating the data collection components and field validating equipment operation. Go to: http://www.itre.ncsu.edu/ncassetmgmtconf/presentations.html
2	Water Bleeding - [Record the number of occurrences of water bleeding and the length in meters of affected pavement with a minimum length of 3.28 feet]	3	2	0	5	0	3	4	4. Optical scanning.; 11. Not sure what this is.
3	Oxidation - [Only plant mix surfaces are rated for oxidation. Determine presence of oxidation]	2	4	0	6	1	2	5	4. Optical scanning.; 11. Not clear what the element is.
4	Bleeding - [Record square feet of surface area at each severity level]	4	1	0	7	1	3	5	4. Optical scanning.
5	Pumping - [Record the number of occurrences of pumping and the length in feet of affected pavement]	2	2	0	7	0	3	5	4. Optical scanning.; 9. Severity and PREVELANCE are both key for impact on safety for EACH of the pavement problems listed
6	Shoving - [Record number of occurrences and square feet of affected surface area]	3	1	0	8	0	3	5	4. Optical scanning.
7	Scaling - [Record the number of occurrences and the square meters of affected area]	3	3	0	6	0	2	6	4. Optical scanning.

#	Data Elements and Categories	A. This element has a high safety priority.	B. Technologies exist to measure this element.	C. Comments
8	Lane-to-Shoulder Separation - [Record to the nearest millimeter at intervals of 50 feet along the lane-to-shoulder joint. Indicate whether the joint is well-sealed (yes or no) at each location]	7 2 0 3	1 3 4	1. Line scanning laser; 4. LIDAR
9	Transverse Construction Joint Deterioration - [Record number of construction joints at each severity level]	5 2 0 4	1 3 4	1. Line scanning laser; 4. LIDAR
	II. Lane Descriptors			
10	Lane Add Point - [GPS or Reference Post of Start of a FULL lane width]	5 4 0 3	2 3 4	4. LIDAR and optical scanning, with appropriate software. (DHMS)
11	Exclusive Left Turn Lane Length - [Exclusive left turn lane length]	8 3 0 1	3 3 3	1. Local DMI or GPS; 4. LIDAR and optical scanning, with appropriate software. (DHMS); 5. Depends on capacity.; 7. Highway Driving Simulator or Field Research Vehicle
12	Auxiliary Lane Presence/Type - [Presence or type of auxiliary lane]	8 1 0 3	1 2 4	4. LIDAR and optical scanning, with appropriate software. (DHMS)
13	Auxiliary Lane Length - [Length of auxiliary lane]	6 2 0 4	1 2 4	4. LIDAR and optical scanning, with appropriate software. (DHMS)
14	HOV Lanes - [Presence of HOV lanes in segment]	6 4 0 2	0 3 5	4. Optical scanning only.; 7. Highway Driving Simulator or Field Research Vehicle
15	HOV Lane Types - [HOV lane types]	5 5 0 2	0 3 5	4. Optical scanning only.; 7. Highway Driving Simulator or Field Research Vehicle

	Data Elements and Categories	A. This element has a high safety priority.				B. Technologies exist to measure this element.			C. Comments
16	Reversible Lanes - [Number of reversible lanes present on segment]	7	2	0	3	0	2	6	4. No current distinguisher between reversible and non-rev. lanes.; 7. Highway Driving Simulator or Field Research Vehicle
17	Presence/Type of Bicycle Facility - [Presence or type of bicycle facility on segment]	11	1	0	0	1	2	5	4. LIDAR and optical scanning, with appropriate software. (DHMS); 7. Highway Driving Simulator or Field Research Vehicle
18	Number of Peak Hour Lanes - [Number of through lanes used in peak period in the peak direction]	8	3	0	1	1	2	5	1. But still good information to collect.; 4. No current distinguisher.; 5. More impact on mobility.; 7. Highway Driving Simulator or Field Research Vehicle
	III. Shoulder, Median, and Roadside Descriptors								
19	Right Shoulder Type - [Shoulder type on right side of road in direction of inventory]	11	1	0	0	1	1	6	4. LIDAR and optical scanning, with appropriate software. (DHMS); 7. Highway Driving Simulator or Field Research Vehicle
20	Right Paved Shoulder Width - [Width of paved portion of right shoulder]	12	0	0	0	1	2	5	4. LIDAR; 7. Highway Driving Simulator or Field Research Vehicle
21	Shoulder Rumble Strip Presence - [Presence of shoulder rumble strip]	11	0	0	1	1	1	6	4. LIDAR and optical scanning, with appropriate software. (DHMS); 7. Highway Driving Simulator or Field Research Vehicle
22	Rumble Strip Type - [Rumble strip type if present]	6	3	0	3	1	1	6	4. LIDAR and optical scanning, with appropriate software. (DHMS);

	Data Elements and Categories	A. This element has a high safety priority.			B. Technologies exist to measure this element.			C. Comments
23	Location of Rumble Strip - Begin - [GPS or reference post of start]	9	1	0	2	1	1	6	7. Field Research Vehicle 4. LIDAR and optical scanning, with appropriate software. (DHMS); 7. Highway Driving Simulator or Field Research Vehicle
24	Rumble Strip Offset - [From edge of lane]	9	1	0	2	1	1	6	4. LIDAR and optical scanning, with appropriate software. (DHMS); 7. Highway Driving Simulator or Field Research Vehicle
25	Sidewalk is separated from edge of road - [Yes/No]	11	1	0	0	1	2	5	1. Video; 4. LIDAR and optical scanning, with appropriate software. (DHMS); 7. Highway Driving Simulator or Field Research Vehicle
26	Curb Type - [Curb type]	6	2	0	4	1	2	5	4. LIDAR and optical scanning, with appropriate software. (DHMS)
27	Curb Location - [GPS coordinates of beginning and end point]	6	4	0	2	2	1	5	4. LIDAR and optical scanning, with appropriate software. (DHMS)
28	Curb Blockage	3	3	0	5	0	2	5	3. I don't know what it is.; 4. Optical scanning. LIDAR if comparing scans.
29	Curb Damage	4	5	0	3	0	3	5	1. Agency may wish to collect the information as part of asset management.; 4. Optical scanning. LIDAR if comparing scans.
30	Median Opening Location - [GPS of first corner in direction of travel]	9	1	0	2	1	2	5	4. LIDAR and optical scanning, with appropriate software. (DHMS)

	Data Elements and Categories	A. This element has a high safety priority.		B. Technologies exist to measure this element.		C. Comments
31	Median Barrier Type - [Soil, paved (striped), paved (barrier), raised curb, None, Other]	12	0	1	2	5	4. LIDAR and optical scanning, with appropriate software. (DHMS); 7. Federal Outdoor Impact Laboratory
32	Barrier beginning point location - [GPS or Reference Post of Beginning of entire barrier system]	9	2	1	2	5	4. LIDAR and optical scanning, with appropriate software. (DHMS); 7. Federal Outdoor Impact Laboratory
33	Barrier offset - beginning - [From edge of lane]	9	2	1	2	5	4. LIDAR and optical scanning, with appropriate software. (DHMS)
34	Barrier height - [Height from ground surface in inches]	11	1	0	3	5	1. LIDAR; 4. LIDAR. Limitation -- identifying ground surface through vegetation, debris, etc.; 7. Federal Outdoor Impact Laboratory
35	Barrier post type - [Strong Post (Metal), Weak Post (Metal), Wooden Post, N/A, Other]	11	1	0	1	6	4. Optical. LIDAR with a library of standard posts.; 7. Federal Outdoor Impact Laboratory
36	Barrier offset bracket - [Yes/No]	4	2	0	1	5	4. LIDAR and optical scanning, with appropriate software. (DHMS)
37	Barrier rub rail - [Yes/No]	5	1	0	1	5	4. LIDAR and optical scanning, with appropriate software. (DHMS)
38	Barrier end treatment type - beginning - [Impact Attenuator, Buried End, Terminal End, Fist, Bridge Connection, None, Other]	10	1	1	1	6	
39	Median Left Turn Lane Type - [Type of left turn lane in median]	10	1	1	3	4	1. Video?; 7. Highway Driving Simulator or Field Research Vehicle
40	Median Left Turn Lane Width - [Width of median left turn lane]	10	1	2	2	4	7. Highway Driving Simulator or Field Research Vehicle

71

	Data Elements and Categories	A. This element has a high safety priority.				B. Technologies exist to measure this element.			C. Comments
41	Drop Inlet Location - [GPS coordinates]	4	2	0	6	1	2	5	
42	Drop Inlet Blockage	4	2	0	6	0	2	5	Optical if inlet in view.
43	Roadside Clearzone Width - [Roadside clearzone width]	11	0	0	0	1	2	5	1. LIDAR and Video?; 4. LIDAR. (With appropriate software.)
44	Roadside Clearzone Cross Slope (Assume Sideslope = Clearzone Cross Slope) - [Sideslope]	8	2	0	2	1	2	5	1. LIDAR and Video?; 4. LIDAR
45	Driveway Type (MMIRE = Driveway Information) - [Residential, Farm, Retail/Commercial, Industrial]	8	3	0	1	0	2	6	
46	Barrier Condition - [Functional or not]	10	0	0	2	0	1	6	
47	Attenuator present (barrier beginning) - [Yes/No]	9	1	0	2	2	1	5	
48	Attenuator type -	7	2	0	3	1	1	6	4. Optical. LIDAR with a library of standard posts.
49	Attenuator Condition - [Functional or not]	9	0	0	3	0	1	5	4. Optical based entirely on operator judgment. LIDAR can generate shape of damaged barrier, but no protocols or technology currently evaluate damage.
50	Offset of roadside obstacle - [From edge of lane]	10	0	0	2	1	3	3	1. LIDAR; 4. LIDAR
51	Location of roadside obstacle - [GPS or reference post of each obstacle]	10	1	0	1	1	3	3	1. LIDAR coupled to GPS; 4. LIDAR or optical.; 7. Highway Driving Simulator or Field Research Vehicle
IV. Bridge Descriptors									
52	Bridge Begin Location (MMIRE: Bridge Descriptors for Bridges in Segment) - [GPS coordinates]	3	4	0	4	1	3	3	1. GPS
53	Bridge approach Slab Settlement Exists - [Yes/No]	6	1	0	4	1	3	3	1. LIDAR; 4. LIDAR.
54	Offset of bridge rail - [Distance from edge of lane]	7	1	0	3	1	3	3	1. Line scanning laser; 4. LIDAR

#	Data Elements and Categories	A. This element has a high safety priority.			B. Technologies exist to measure this element.			C. Comments
55	Bridge Median - [Whether the median is non-existent, open or closed. The median is closed when the area between the 2 roadways at the structure is bridged over and is capable of supporting traffic. All bridges that carry either 1-way traffic or 2-way traffic separated only by a centerline have no median]	9	0	0	2	1	3	1. Line scanning laser; 4. Optical or LIDAR.
56	Bridge transitions - [The transition from approach guardrail to bridge railing requires that the approach guardrail be firmly attached to the bridge railing. It also requires that the approach guardrail be gradually stiffened as it comes closer to the bridge railing. The ends of curbs and safety walks need to be gradually tapered out or shielded]	7	2	0	2	0	1	6
57	Type of Service on Bridge - [Classification: Highway, Railroad, Pedestrian-bicycle, Highway-railroad, Highway-pedestrian, Overpass structure at an interchange or second level of a multilevel interchange, Third level (Interchange), Fourth level (Interchange), Building or plaza, Other]	7	1	0	3	0	2	5	1. Existing inventory
58	Left Curb/Sidewalk Width - [Widths of the left and right curbs or sidewalks to nearest tenth of a meter. "Left" and "Right" should be determined on the basis of direction of the inventory]	8	1	0	2	1	2	3	1. Line scanning laser; 4. LIDAR

V. Railroad Crossing Descriptors

#	Data Elements and Categories	A. This element has a high safety priority.			B. Technologies exist to measure this element.			C. Comments
59	Number of Tracks - [Number of railroad tracks]	8	2	0	2	1	2	5	1. LIDAR and Google Earth; 4. Optical or LIDAR.
60	Railroad Crossing Control Type - [Crossbucks, gates, flashing lights, signal]	10	1	0	1	1	1	6	1. Video; 4. Optical. LIDAR with a library of standard features.
61	Railroad crossing number - [U.S. DOT inventory crossing number]	6	3	0	3	0	1	6	1. Existing inventory

	Data Elements and Categories	A. This element has a high safety priority.			B. Technologies exist to measure this element.		C. Comments
62	Railroad crossing Four-quadrant (or Full Barrier) Gates - [Whether or not four-quadrant (or full barrier) gates are present at the crossing. Full barrier gates apply in the case of 1-way streets or where the gate arms reach across the entire roadway]	9	0	0	3	0	2	6	1. Video
63	Railroad crossing Cantilevered (or Bridged) Flashing Lights - [The number of cantilevered (or bridged) flashing lights in the appropriate block. Separate cantilevered flashers from those over traffic lanes and those not reaching the roadway (over only parking lanes, turnout lanes, or shoulders). Count individual cantilever units; do not count the flasher head pairs mounted on the units]	7	1	0	4	1	1	5	1. Video; 4. Optical, or LIDAR with library of elements.
64	Railroad crossing Wigwags - [The number of wigwag signals]	7	1	0	4	1	1	6	1. Video?; 4. Optical, or LIDAR with library of elements.
65	Railroad crossing Bells - [The number of all bells, if present, that are either alone or in conjunction with other train activated warning devices]	6	1	0	4	0	1	7	5. Bells are difficult to hear inside car.
66	Crossing Surface (on Main Line) - [Type of crossing surface (if there are multiple tracks which have different types of surfaces, indicate the lower grade surface material): Timber, Sectional Treated Timber, Full Wood Plank, Asphalt, Asphalt and Flange, Concrete, Concrete Slab, Concrete and Rubber, Rubber, Metal, Metal Sections, Other Metal, Unconsolidated, Other]	6	1	0	4	0	2	5	4. Optical only.

VI. Traffic Operations/Control Descriptors

67	One/Two-Way Operations - [Whether the segment operates as a one- or two-way roadway]	10	1	0	0	0	2	6	7. Highway Driving Simulator or Field Research Vehicle
68	On-Street Parking Type - [On-street parking type]	8	2	0	1	1	2	5	1. Google Earth?; 4. Optical

	Data Elements and Categories	A. This element has a high safety priority.				B. Technologies exist to measure this element.			C. Comments
69	Beginning of on-street parking - [GPS or Reference of start]	9	1	0	2	1	1	6	4. Optical or LIDAR.
70	Roadway Lighting Presence - [Yes/No]	12	0	0	0	1	3	4	1. Vehicle mounted photometer.; 4. Optical.; 7. Field Research Vehicle
71	Location of street lighting - [GPS or reference post of pole]	11	1	0	0	2	2	4	1. LIDAR and photometer.; 4. Optical or LIDAR.; 5. DHM with video cross ref.?; 7. Field Research Vehicle
72	Toll Facility? - [Toll facility indicator]	7	4	0	1	0	2	6	7. Highway Driving Simulator or Field Research Vehicle
73	Edgeline Marking Width - [To the nearest inch]	10	1	0	1	1	3	4	1. Mobile Retroreflectometer.; 4. LIDAR.; 7. Highway Driving Simulator or Field Research Vehicle
74	Edgeline Marking Color -	6	1	0	5	1	2	5	4. Optical.
75	Edgeline Location of marking - begin - [GPS or Reference of start]	11	1	0	0	2	2	4	1. GPS.; 4. Optical or LIDAR.; 7. Highway Driving Simulator or Field Research Vehicle
76	Edgeline Marking Material Type - [Paint, Thermoplastic, Polyurethane]	6	?	0	4	0	2	6	7. Highway Driving Simulator or Field Research Vehicle

	Data Elements and Categories	A. This element has a high safety priority.			B. Technologies exist to measure this element.		C. Comments
77	Edgeline Marking offset - [Offset of each type of line (center, lane & edge) from right edge of pavement]	9	2	0	1	2	4	1. Provide calculated lane width? Use mobile pavement marking retroreflectometer to determine line separation.; 4. LIDAR. Optical can come up with a measure, but tolerances fairly large.
78	Edgeline marking type	3	2	0	6	1	5	
VII. Horizontal and Vertical Curve Descriptors								
79	Curve Feature Type - [Type of horizontal alignment feature being described in the data record]	9	1	0	2	1	2	4	1. Google Earth?
80	Sight Distance (Stopping) - [Report sight distance at 0.01-mile intervals for Horizontal Curves from PC to PT]	10	1	0	1	2	2	4	1. LIDAR; 4. Optical. (Potentially LIDAR point cloud can be processed to calculate sight distance.); 7. Field Research Vehicle
81	Sight Distance (Stopping) - [Report sight distance at 0.01-mile intervals for Vertical Curves from PC to PT]	11	0	0	0	2	2	4	1. LIDAR; 4. Optical. (Potentially LIDAR point cloud can be processed to calculate sight distance.); 7. Field Research Vehicle
VIII. Intersection and Interchange Descriptors								
82	Type of Intersection/Junction - [Type of junction being described in the data record]	11	1	0	0	1	3	4	4. Optical.; 7. Highway Driving Simulator or Field Research Vehicle
83	Location Identifier for Road 1 Crossing Point - [Location on the first intersecting route (e.g., route-milepost)]	7	2	0	3	1	1	6	7. Highway Driving Simulator or Field Research Vehicle
84	Intersection/Junction No. of Legs - [Intersection/junction no. of legs]	10	1	0	1	2	1	5	1. Google Earth?; 4. Optical or LIDAR.

	Data Elements and Categories	A. This element has a high safety priority.			B. Technologies exist to measure this element.		C. Comments
85	Intersection/Junction Geometry - [Intersection/junction geometry]	11	1	0	2	1	1. Google Earth?; 4. LIDAR with appropriate software.; 7. Highway Driving Simulator or Field Research Vehicle
86	Intersection Skew Angle - [Angle from perpendicular of intersection of the roads]	11	1	0	0	3	1. Google Earth?; 4. LIDAR.; 7. Highway Driving Simulator or Field Research Vehicle
87	Intersection/Junction Offset - [Whether crossroad approach centerlines are directly opposed or offset by some distance]	8	2	2	0	3	1. Google Earth?; 4. LIDAR
88	Intersection/Junction Offset Distance - [Distance that approach centerlines are offset]	6	2	4	0	3	1. Google Earth?
89	Type of signalized intersection - [Standard, Protected Turn, Permitted Turn]	11	1	0	0	1	7. Highway Driving Simulator or Field Research Vehicle
90	Flashing beacon present [at stop-controlled intersection] - [Yes/No (Flashing yellow/red beacon)]	10	2	0	1	1	4. Optical.; 7. Highway Driving Simulator or Field Research Vehicle
91	Number of Intersection/Junction Quadrants With Limited Sight Distance - [Number of intersection/junction quadrants with limited sight distance]	11	1	0	0	2	7. Highway Driving Simulator or Field Research Vehicle
92	Roundabout-Inscribed Diameter - [Distance between the outer edges of the circulatory roadway]	9	1	2	1	3	4. LIDAR.; 7. Highway Driving Simulator
93	Channelization exists on approach - [Yes/No]	10	1	1	2	1	4. Optical or LIDAR.; 7. Highway Driving Simulator or Field Research Vehicle
94	Pedestrian Signalization Type - [Type of pedestrian signalization on approach]	10	1	1	0	2	4. Optical

	Data Elements and Categories	A. This element has a high safety priority.		B. Technologies exist to measure this element.		C. Comments
95	Roundabout-Entry Radius - [Minimum radius of curvature of the curb on the right side of the entry]	9	1 0 2	2	2 4	4. LIDAR; 7. Highway Driving Simulator or Field Research Vehicle
96	Roundabout-Pedestrian Facility - [Type of pedestrian crossing facility on this approach to roundabout]	9	1 0 1	1	2 5	1. Video?; 4. Optical.; 7. Highway Driving Simulator or Field Research Vehicle
97	Roundabout-Splitter Island Width - [Width of the splitter island separating entry and exit legs (measured at the inscribed circle)]	8	1 0 3	2	1 5	4. LIDAR; 7. Highway Driving Simulator or Field Research Vehicle
98	Ramp Location - [GPS Coordinates or Reference Post of gore area]	8	2 0 2	1	4 3	1. GPS; 4. Optical or LIDAR.
99	Ramp Length - [Length of ramp]	10	1 0 1	1	4 3	1. Local DMI or GPS; 4. Optical or LIDAR.; 5. Depends on speed.; 7. Highway Driving Simulator or Field Research Vehicle
100	Location Identifier For Roadway at Beginning Ramp Terminal - [Location on the roadway at the beginning ramp terminal (e.g., route-milepost for that roadway) if the ramp connects with a roadway at that point]	6	3 0 3	1	2 5	4. Optical or LIDAR.

Appendix D: FHWA List of Element Priorities

The element priorities provided by FHWA as well as the background and criteria for their classification[12] are presented in the following sections.

D.1 Background

Most agencies with responsibilities for the safety of road and highway systems clearly recognize the need for comprehensive, high-quality data about their roads and roadside environments. Nationally, this is reflected in FHWA's strategic commitment to data-driven decisions.[13]

A number of projects and programs are currently underway to identify the information needed to understand what is happening on U.S. roads and highways, and to provide analysts and designers with the data which will allow them to identify, characterize and prioritize safety improvements to best use the limited funds available for making our roads safer. These projects include:

- Safety analysis tools: The Interactive Highway Safety Design Model (IHSDM) and SafetyAnalyst are two recently-developed analysis tools which use data about a road system and its operational details to help practitioners make good safety decisions.
- Minimum Inventory of Roadway Elements (MIRE – formerly MMIRE): MIRE is a list of standard definitions of important data elements to collect about a road system, along with standard forms for coding that data for inclusion in databases.

---

[12] The information was provided by Mr. Lincoln Cobb, FHWA Office of Safety R&D, HRDS-2.
[13] FHWA Strategic Objectives – System Performance Goal 2.1: *"Implement comprehensive, integrated, and data-driven safety programs and countermeasures at the Federal, State, and local level."*

- Strategic Highway Research Program 2 (SHRP2): The Safety Focus area of SHRP2 will collect enormous data sets regarding the roads, and their immediate environments, travelled by volunteer subjects in the Naturalistic Driving Study.
- Intellidrive Applications: A current trade study is considering the roadway data needs for Intellidrive applications. Intellidrive data needs are not precisely those of a highway safety analyst, but there are considerable overlaps. Intellidrive systems will need to distinguish between more and less safe conditions through which the vehicle is passing.

The roadway safety community has conducted little research which would allow objective ranking of the potential impact of safety data elements. When asked which roadway data elements are most important, experts often respond by gut feel. What has been and remains the case is that everyone involved in roadway safety has an opinion about what information is important, and each may even have devolved a personal ranking of the elements. Therefore, the use of road data elements in contemporary safety analysis software tools (whose data elements have a practical impact on safety) was chosen as a surrogate for the direct significance of individual elements.

D.2 Criteria for Defining High Priority Data Elements

Among the data elements which would support the best safety decisions, the *most common* denominator is geometric alignment information, particularly basic information about horizontal and vertical curvature. All of the above-mentioned data sets or sources include such basic geometric elements. In each instance this basic geometric information is considered critical though not typically ranked higher or lower than other "important" data elements.

In selecting the specific elements to include in the High Priority list, the alignment elements from several sources were reviewed:

- MIRE
- SHRP2 S04 project (in progress).
- Interim data element set proposed within the ongoing "Roadway Geometry and Inventory Trade Study for IntelliDriveSM Applications."

This review found no new underlying alignment characteristics, though different data sets reflected different notions about the minimum information needed regarding vertical and horizontal curves. Accordingly, the existing elements in the Roadway Alignment Descriptors section of the MET were defined as the High Priority elements (see Table D.1).

Table D.1. List of FHWA High Priority Elements

MET Element Index	Element Description
297	Curve Identifiers and Linkage Variables
298	Curve Feature Type
299	Horizontal Curve Degree or Radius
300	Horizontal Curve Length (Including Spiral)
301	Curve Superelevation or Superelevation Adequacy
302	Horizontal Curve PC (Point of Curvature)
303	Horizontal Curve PT (Point of Tangency)
306	Horizontal Curve Intersection/Deflection Angle
307	Horizontal Curve Direction
308	(Horizontal) Sight Distance (Stopping)
309	Grade Identifiers and Linkage Variables
310	Vertical Alignment Feature Type
311	Percent of Grade
312	Location of measurement [of grade]
313	Grade Length
314	Vertical Curve Length
315	Vertical Curve Radius
317	Vertical Curve PC (Point of Curvature)
318	Vertical Curve PT (Point of Tangency)
319	(Vertical) Sight Distance (Stopping)

D.3 Criteria for Defining Medium Priority Data Elements

There are two safety analysis tools in particular – IHSDM and SafetyAnalyst – which have taken prominent positions in the road safety field. These are fundamental parts of the first release of the Highway Safety Manual (HSM), which will be published in the late Spring of 2010. Therefore, the road data elements which are inputs for these two tools will have considerable impact on safety via the analyses which will be conducted using them. On this basis, the inputs into IHSDM and SafetyAnalyst are collectively considered the Medium Priority Data Elements for the purposes of this study and are presented in Table D.2.

Table D.2. List of FHWA Medium Priority Elements

MET Element Index	Element Description
1	County
2	City/Local Jurisdiction
3	Route Number
5	Section End-Points Descriptors
6	Section Identifier
7	Section Length
8	Highway District
11	Route Signing
16	Functional Class
17	Rural/Urban Designation
19	Access Control
21	Surface Type
63	No. of Thru Lanes
65	Lane Widths
69	Average Thru Lane Width
76	HOV Lanes
79	Presence/Type of Bicycle Facility
83	Right Shoulder Type
84	Right Shoulder Total Width
85	Right Paved Shoulder Width
90	Left Shoulder Type

MET Element Index	Element Description
91	Left Shoulder Total Width
92	Left Paved Shoulder Width
97	Shoulder Rumble Strip Presence
112	Median Type
113	Median Width
116	Median Barrier Type
136	Median Left Turn Lane Type
138	Roadway Cross Slope
139	Location of measurement [of Cross Slope]
146	Sideslope
148	Roadside Rating
150	Driveway Information
150	Driveway Type (MMIRE = Driveway Information)
174	Type of roadside obstacles
175	Offset of roadside obstacle
176	Location of roadside obstacle
177	Terrain Type (e.g., Mountainous, Level)
242	Average Annual Daily Traffic Volume
243	AADT Year
244	AADT Annual Escalation Percentage
245	Percentage Truck or Truck AADT
247	Bicycle Count/Exposure
249	Hourly Traffic Volumes (or Peak and Off-Peak
251	Future AADT
252	Future AADT Year
253	Directional Factor
256	One/Two-Way Operations
257	Speed Limit
259	On-Street Parking Presence
260	On-Street Parking Type
261	Beginning of on-street parking
262	Ending of on-street parking
263	Side of street with on-street parking
264	Roadway Lighting
265	Roadway Lighting Presence
296	85th % Speed
297	Curve Identifiers and Linkage Variables
299	Horizontal Curve Degree or Radius
300	Horizontal Curve Length (Including Spiral)
301	Curve Superelevation or Superelevation Adequacy
304	Horizontal Transition/Spiral Curve Presence
307	Horizontal Curve Direction

MET Element Index	Element Description
311	Percent of Gradient
313	Grade Length
320	Unique Intersection Identifier
322	Location Identifier for Road 1 Crossing Point
324	Location Identifier for Road 3, 4, etc., Crossing
325	Intersection/Junction No. of Legs
326	Intersection/Junction Geometry
329	Intersection Skew Angle
330	Intersection/Junction Offset
331	Intersection/Junction Offset Distance
332	Intersection/Junction Traffic Control
333	Signalization Type (e.g. Actuated, Fixed, System)
337	Number of Intersection/Junction Quadrants With
338	Intersection/Junction Lighting
344	Approach AADT
346	Approach is Two-Way, One-Way
349	No. of Exclusive Left Turn Lanes
351	No. of Exclusive Right Turn Lanes
354	Median Type at Intersection
355	Approach Traffic Control
357	Left Turn Protection
358	Signal Progression
362	Crossing Pedestrian Count/Exposure
363	Left/Right Turn Prohibitions
364	Left Turn Counts/Percent
365	Right Turn Counts/Percent
380	Interchange Type
383	Ramp Location
384	Type of Ramp Terminal [entry or exit]
385	Ramp Length
386	Ramp No. of Lanes
387	Ramp AADT
390	Ramp Descriptor at Beginning Ramp Terminal
394	Ramp Descriptor at Ending Ramp Terminal

D.4 Criteria for Defining Low Priority Data Elements

The data elements in the MET that are not classified as High or Medium are considered Low Priority Data Elements.

Appendix E: Combined List of Element Priorities

Tables E.1, E.2 and E.3 present the high, medium and low, respectively, prioritized list of data elements that were merged from the FHWA survey (Appendix C) and the FHWA lists given in Appendix D.

Table E.1. High Priority Data Elements

MET Element Index	Element Description
34	Polished Aggregate
40	Bleeding
46	Shoving
54	Lane-to-Shoulder Separation
55	Transverse Construction Joint Deterioration
71	Exclusive Left Turn Lane Length
74	Auxiliary Lane Presence/Type
75	Auxiliary Lane Length
78	Reversible Lanes
79	Presence/Type of Bicycle Facility
82	Number of Peak Hour Lanes
83	Right Shoulder Type
85	Right Paved Shoulder Width
97	Shoulder Rumble Strip Presence
98	Rumble Strip Type
100	Location of Rumble Strip - Begin
102	Rumble Strip Offset
106	Sidewalk is separated from edge of road
108	Curb Type
115	Median Opening Location
116	Median Barrier Type
120	Barrier offset - beginning
122	Barrier height
123	Post type
124	Offset bracket
125	Rub rail
133	Location of Rumble Strip - Begin
135	Rumble Strip Offset From edge of lane
136	Median Left Turn Lane Type
137	Median Left Turn Lane Width
143	Drop Inlet Location
144	Drop Inlet Blockage
145	Roadside Clearzone Width
146	Roadside Clearzone Cross Slope (Assume Sideslope = Clearzone Cross Slope)
150	Driveway Type (MMIRE = Driveway Information)
153	Barrier Begin Location

MET Element Index	Element Description
155	Barrier offset - beginning
158	Post type
159	Offset bracket
160	Rub rail
161	Barrier Condition
162	Barrier end treatment (beginning)
164	Attenuator present (barrier beginning)
165	Attenuator type
166	Attenuator Condition
175	Offset of roadside obstacle
176	Location of roadside obstacle
180	Approach Slab Settlement Exists
182	Offset of bridge rail
186	Bridge Median
188	Transitions
191	Type of Service on Bridge
194	Left Curb/Sidewalk Width
202	Number of Tracks
203	Railroad Crossing Control Type
206	Railroad crossing number
219	Four-quadrant (or Full Barrier) Gates
220	Cantilevered (or Bridged) Flashing Lights
225	Wigwags
226	Bells
234	Crossing Surface (on Main Line)
256	One/Two-Way Operations
260	On-Street Parking Type
261	Beginning of on-street parking
265	Roadway Lighting Presence
266	Location of street lighting
270	Edgeline Marking Width
271	Edgeline Marking Color
272	Edgeline Location of marking - begin
274	Edgeline Marking Material Type
275	Edgeline Marking offset
297	Curve Identifiers and Linkage Variables
298	Curve Feature Type
299	Horizontal Curve Degree or Radius
300	Horizontal Curve Length (Including Spiral)
301	Curve Superelevation or Superelevation Adequacy
302	Horizontal Curve PC (Point of Curvature)
303	Horizontal Curve PT (Point of Tangency)
306	Horizontal Curve Intersection/Deflection Angle
307	Horizontal Curve Direction
308	Sight Distance (Stopping)
309	Grade Identifiers and Linkage Variables
310	Vertical Alignment Feature Type
311	Percent of Gradient
312	Location of measurement [of grade]
313	Grade Length

MET Element Index	Element Description
314	Vertical Curve Length
315	Vertical Curve Radius
317	Vertical Curve PC (Point of Curvature)
318	Vertical Curve PT (Point of Tangency)
319	Sight Distance (Stopping)
321	Type of Intersection/Junction
322	Location Identifier for Road 1 Crossing Point
325	Intersection/Junction No. of Legs
326	Intersection/Junction Geometry
329	Intersection Skew Angle
330	Intersection/Junction Offset
331	Intersection/Junction Offset Distance
334	Type of signalized intersection
336	Flashing beacon present [at stop-controlled intersection]
337	Number of Intersection/Junction Quadrants With Limited Sight Distance
341	Roundabout-Inscribed Diameter
343	Channelization exists on approach
360	Pedestrian Signalization Type
369	Roundabout-Entry Radius
373	Roundabout-Pedestrian Facility
375	Roundabout-Splitter Island Width
383	Ramp Location
385	Ramp Length
391	Location Identifier For Roadway at Beginning Ramp Terminal

Table E.2. Medium Priority Data Elements

MET Element Index	Element Description
1	County
2	City/Local Jurisdiction
3	Route Number
5	Section End-Points Descriptors
6	Section Identifier
7	Section Length
8	Highway District
11	Route Signing
16	Functional Class
17	Rural/Urban Designation
19	Access Control
21	Surface Type
41	Pumping
52	Scaling
63	No. of Thru Lanes
65	Lane Widths
67	Lane Add Point
69	Average Thru Lane Width

MET Element Index	Element Description
76	HOV Lanes
77	HOV Lane Types
84	Right Shoulder Total Width
90	Left Shoulder Type
91	Left Shoulder Total Width
92	Left Paved Shoulder Width
109	Curb Location
110	Curb Blockage
111	Curb Damage
112	Median Type
113	Median Width
138	Roadway Cross Slope
139	Location of measurement [of Cross Slope]
148	Roadside Rating
174	Type of roadside obstacles
177	Terrain Type (e.g., Mountainous, Level)
178	Bridge Begin Location (MMIRE: Bridge Descriptors for Bridges in Segment)
242	Average Daily Traffic Volume
243	AADT Year
244	AADT Annual Escalation Percentage
245	Percentage Truck or Truck AADT
247	Bicycle Count/Exposure
249	Hourly Traffic Volumes (or Peak and Off-Peak AADT)
251	Future AADT
252	Future AADT Year
253	Directional Factor
257	Speed Limit
259	On-Street Parking Presence
262	Ending of on-street parking
263	Side of street with on-street parking
264	Roadway Lighting
268	Toll Facility?
277	Edgeline marking type
296	85th % Speed
304	Horizontal Transition/Spiral Curve Presence
320	Unique Intersection Identifier
324	Location Identifier for Road 3, 4, etc., Crossing Point (e.g., Route-Milepost), etc.
332	Intersection/Junction Traffic Control
333	Signalization Type (e.g., Actuated, Fixed, System)
338	Intersection/Junction Lighting
344	Approach AADT
346	Approach Is Two-Way, One- Way
349	No. of Exclusive Left Turn Lanes
351	No. of Exclusive Right Turn Lanes
354	Median Type at Intersection
355	Approach Traffic Control
357	Left Turn Protection
358	Signal Progression
362	Crossing Pedestrian Count/Exposure
363	Left/Right Turn Prohibitions

MET Element Index	Element Description
364	Left Turn Counts/Percent
365	Right Turn Counts/Percent
380	Interchange Type
384	Type of Ramp Terminal [entry or exit]
386	Ramp No. of Lanes
387	Ramp AADT
390	Ramp Descriptor at Beginning Ramp Terminal
394	Ramp Descriptor at Ending Ramp Terminal

Table E.3. Low Priority Data Elements

MET Element Index	Element Description
4	Street Name
9	Governmental Ownership
10	Type of Governmental Ownership
12	Route Signing Qualifier
13	Coinciding Route Indicator
14	Coinciding-Route Primary Route Number
15	Direction of Inventory
18	Federal Aid/ Route Type
20	Operational Class
22	Surface Friction
23	Surface Texture
24	Surface Friction Date
25	Total Surface Width
26	Roughness
27	Roughness Date
28	Condition
29	Condition Date
30	Critical Failure
31	Patches (type & no. & area)
32	Longitudinal Cracking
33	Transverse Cracking
35	Lane-to-Shoulder Drop-off
36	Water Bleeding
37	Alligator Cracking (Fatigue Cracking) [area]
38	Raveling
39	Oxidation
42	Block Cracking
43	Edge Cracking
44	Reflection Cracking at Joint
45	Potholes
47	Rutting
48	Durability Cracking (D-Cracking)
49	Map Cracking
50	Spalling of longitudinal joint

MET Element Index	Element Description
51	Lane-Joint Seal Damage
53	Blow up
56	Corner Cracking (Corner Break)
57	Spalling of transverse joint
58	Joint Seal Damage
59	Faulting of Transverse Joints
60	Punchouts
61	Narrow Cracks
62	'Y' Cracks
64	Number of Lanes
66	Location of Measurement [of lane widths]
68	Lane Drop Point
70	Exclusive Left Turn Lane Presence
72	Exclusive Right Turn Lane Presence
73	Exclusive Right Turn Lane Length
80	Width of Marked Bicycle Lane or Bike Path
81	Width of Wide Curb Lane
86	Right Location of Measurement [of width]
87	Right Low Shoulder
88	Right High Shoulder
89	Right Lane-joint seal type
93	Left Location of Measurement [of width]
94	Left Low Shoulder
95	Left High Shoulder
96	Left Lane-joint seal type
99	Rumble Strip Lateral Location
101	Location of Rumble Strip - End
103	Sidewalk Presence
104	Location of Sidewalk - Begin
105	Location of Sidewalk - End
107	Curb Presence
114	Location of Measurement [of width]
117	Location
118	Beginning point location
119	Ending point location
121	Barrier offset - ending
126	End treatment type - beginning
127	End treatment type - end
128	Median (Inner) Paved Shoulder Width
129	Median Shoulder Rumble Strip Presence
130	Median Rumble Strip Type
132	Rumble Strip Lateral Location
134	Location of Rumble Strip - End
141	Distance of Pavement Edge Drop-off
142	Location of Measurement [of drop-off]
147	Location of measurement [of Cross Slope]
149	Driveway Location (MMIRE = Driveway Information)
151	Barrier Type (MMIRE Roadside Hardware Descriptors breaks up into all elements left and below)
152	Barrier Location

MET Element Index	Element Description
154	Barrier End Location
156	Barrier offset - ending
157	Barrier height
163	Barrier end treatment (end)
167	Sign Location
168	[Sign] Support Type
169	[Sign] Support Location
170	[Sign] Multi-sign
171	Sign Type(s)
172	Sign Size
173	Sign Retroreflectivity
179	Bridge End Location
181	Bridge Rail Exists
183	Number of lanes on structure
184	Number of lanes under structure
185	Approach Roadway Width
187	Bridge Railings
189	Approach Guardrail
190	Approach Guardrail Ends
192	Type of Service Under Bridge
193	Inventory Rte Total Horz Clearance
195	Right Curb/Sidewalk Width
196	Bridge Roadway Width Curb-to-Curb
197	Min Vert Clear Over Bridge Roadway
198	Minimum Vertical Underclearance
199	Minimum Lateral Underclearance On Right
200	Min Lateral Underclearance on Left
201	At-Grade Railroad Crossing Location (RR Grade Crossing Descriptors for Crossings in Segment)
204	Grade of Approach Side of Crossing
205	Grade of Leave Side of Crossing
207	Crossing Position
208	Latitude (Decimal)
209	Longitude (Decimal)
210	No Signs or Signals
211	Crossbucks
212	Highway Stop Signs (R1-1)
213	RR Advance Warning Signs (W10-1)
214	Hump Crossing Sign (W10-5)
215	Pavement Markings
216	Other Signs
217	Type of Warning Device at Crossing - Signs
218	Gates
221	Mast Mounted Flashing Lights
222	Number of Flashing Light Pairs
223	Other Flashing Lights
224	Highway Traffic Signals
227	Other Train Activated Warning Devices
228	Specify Special Warning Device NOT Train Activated
229	Channelization Devices With Gates

MET Element Index	Element Description
230	Smallest Crossing Angle
231	Number of Traffic Lanes Crossing Railroad
232	Are Truck Pullout Lanes Present?
233	Is Highway Paved?
235	Does Track Run Down a Street?
236	Nearby Intersecting Highway?
237	Is It Signalized?
238	Is Crossing Illuminated?
239	Number of Signalized Intersections in Section
240	Number of Stop-Controlled Intersections in Section
241	Number of Uncontrolled/Other Intersections
246	Total Daily Two-Way Pedestrian Count/Exposure
248	Motorcycle Count or Percentage
250	K-Factor
254	Percent Combination Trucks - Daily Average
255	Percent Single Unit Trucks - Daily Average
258	School Zone Indicator
267	Truck Route Designation
269	Edgeline Presence
273	Edgeline Location of marking - end
276	Edgeline Marking retroreflectivity
279	Centerline Presence
280	Centerline Marking Width
281	Centerline Marking Color
282	Centerline Location of marking - begin
283	Centerline Location of marking - end
284	Centerline Bearing
285	Centerline Marking Material Type
286	Centerline Marking offset
287	Centerline Marking retroreflectivity
288	Centerline marking type
289	Special pavement marking location
290	Special pavement marking description
291	Special Marking Material Type
292	Raised pavement markers present
293	Raised Pavement Marking Type
294	Location of raised pavement markers
295	No Passing Zone Code / Passing Permissibility
305	Transition Curve Length
323	Location Identifier for Road 2 Crossing Point
327	School Zone Indicator
328	Railroad Crossing Number if a RR Grade Crossing
335	Type of stop-controlled intersection
339	Roundabout - No. of Circulatory Lanes
340	Roundabout - Circulatory Width
342	Roundabout-Bicycle Facility
345	Approach Use Type
347	No. of Thru Lanes
348	Exclusive left turn lanes exist
350	Exclusive right turn lanes exist

MET Element Index	Element Description
352	Length of Exclusive Left Turn Lanes
353	Length of Exclusive Right Turn Lanes
356	Location of traffic signal poles
359	Crosswalk Presence/Type
361	Pedestrian Signal Special Features
366	Transverse Rumble Strip Presence
367	Roundabout-Entry Width
368	Roundabout-Number of Entry Lanes
370	Roundabout-Exit Width
371	Roundabout-Number of Exit Lanes
372	Roundabout-Exit Radius
374	Roundabout-Crosswalk Location (Distance From Yield Line)
376	Unique Interchange Identifier
377	Location Identifier for Road 1 Crossing Point
378	Location Identifier for Road 2 Crossing Point
379	Location Identifier for Road 3, 4, etc., Crossing Point, etc.
381	Interchange Lighting
382	Unique Ramp Identifier
388	Ramp Posted Speed Limit
389	Feature at Beginning Ramp Terminal
392	Roadway Traffic Flow Direction at Beginning Ramp Terminal
393	Feature at Ending Ramp Terminal
395	Location Identifier for Roadway at Ending Ramp Terminal
396	Roadway Traffic Flow Direction at Ending Ramp Terminal

Bibliography

3D laser mapping (2009). StreetMapper Products, from
http://www.3dlasermapping.com/uk/mobile/streetmapper/technology.htm

AASHTO (2004). *A Policy on Geometric Design of Highway and Streets* (Fifth Edition ed.): AASHTO.

AASHTO (2005a). Standard Practice for Measuring Pavement Profile Using a Dipstick, *PP41-05*.

AASHTO (2005b). Standard Practice for Measuring Pavement Profile Using a Rod and Level, *PP40-05*.

AASHTO (2005c). Standard Practice for Quantifying Cracks in Asphalt Pavement Surface, *PP44-01*.

AASHTO (2007). Standard Practice for Quantifying Roughness of Pavements, *R43M / R43-07*.

AASHTO (2008). Standard Practice for Determining Rut Depth in Pavements, *PP48-08*.

AASHTO-AGC-ARTBA Joint Committee - Subcommittee on New Highway Materials (10/12/05). A Guide to Standardized Highway Barrier Hardware, Online Hardware Guide - System Index. *Task Force 13 Report*, from http://aashtotf13.tamu.edu/Guide/nameindex.html#grails

Acuity Laser Measurement (2009). High speed, longitudinal road profiling, from
http://www.acuitylaser.com/AR700/common-applications-road-profiling-sensor.shtml

Administration, Federal Railroad. Assignment of Crossing Inventory Numbers Retrieved June 24, 2009, from http://www.fra.dot.gov/downloads/safety/AssignmentofNumbers2006.pdf

Äijö, Juha (2005). *Automated Crack Measurement Test in Finland 2004*. Helsinki: Finnish Road Administration.

Alefs, Bram, Eschemann, Guy, Ramoser, Herbert, & Beleznai, Csaba (2007). *Road Sign Detection from Edge Orientation Histograms*. Paper presented at the Proceedings of the 2007 IEEE Intelligent Vehicles Symposium, Istanbul, Turkey.

ambercore (2009). Titan - Mobile Laser Scanning, from http://www.ambercore.com/index.php

Andrey, Vavilin, & Jo, Kang Hyun (2006). *Automatic Detection and Recognition of Traffic Signs using Geometric Structure Analysis*. Paper presented at the SICE-ICASE International Joint Conference. from http://203.250.84.120/new/papers/30andy1.pdf

Applanix (2009). LAND Mark, from
http://www.applanix.com/index.php?option=com_content&view=article&id=124&Itemid=79

ARRB Group (2009). Advancing safety and efficiency in transport through knowledge, from www.arrb.com.au

ASFT (2009). Technology, from http://www.asft.se/road/Technology.shtml

ASTM (2004). Standard Test Method for Measuring the Longitudinal Profile of Traveled Surfaces with an Accelerometer Established Inertial Profiling Reference, *E950-98*.

ASTM (2005a). Standard Test Method for Calculating Pavement Macrotexture Mean Profile Depth, *E1845-01*.

ASTM (2005b). Standard Test Method for Measuring Road Roughness by Static Level Method, *E1364-95*.

ASTM (2006a). Standard Test Method for Friction Coefficient Measurements Between Tire and Pavement Using a Variable Slip Technique, *E1859-97*.

ASTM (2006b). Standard Test Method for Measuring Pavement Macrotexture Depth Using a Volumetric Technique, *E965-96(2006)*.

ASTM (2006c). Standard Test Method for Skid Resistance of Paved Surfaces Using a Full-Scale Tire, *E274-06*.

ASTM (2008a). Standard Test Method for Computing International Roughness Index of Roads from Longitudinal Profile Measurements, *E1926-08*.

ASTM (2008b). Standard Test Method for Determining Longitudinal Peak Braking Coefficient of Paved Surfaces Using Standard Reference Test Tire, *E1337-90*.

ASTM (2008c). Standard Test Method for Measuring Pavement Roughness Using a Profilograph, *E1274-03*.

ASTM (2009). Standard Terminology for Three-Dimensional (3D) Imaging Systems: ASTM.

Aufrere, Romuald, Mertz, Christoph, & Thorpe, Charles (2003). *Multiple Sensor Fusion for Detecting Location of Curbs, Walls, and Barriers*. Paper presented at the Proceedings of the IEEE Intelligent Vehicles Symposium.

Bahar, Geni, Mollett, Calvin, Persaud, Bhagwant, Lyon, Craig, Smiley, Alison, Smahel, Tom, et al. (2004). *Safety Evaluation of Permanent Raised Pavement Markers* (No. NCHRP Report 518).

Bay, J. A., & Stokoe, K. H. (1998). *Development of a Rolling Dynamic Deflectometer for Continuous Deflection Testing of Pavements* (No. 1422-3F).

Blincoe, L., Seay, A., Zaloshnja, E., Miller, T., Romano, E., Luchte, S., et al. (2002). *The Economic Impact of Motor Vehicle Crashes, 2000*. Washington, DC: National Highway Traffic Safety Administration, U.S. Department of Transportation.

Canedy, Thomas (1997). USA Patent 5701122 Patent No. 5701122.

Carlson, Paul J., Burris, Mark, Black, Kit, & Rose, Elisabeth R. (2005). Comparison of Radius-Estimating Techniques for Horizontal Curves. *Transportation Research Record: Journal of the Transportation Research Board, 1918*, 76 - 83.

Central Federal Lands Highway. Reversible Lanes, from http://www.cflhd.gov/ttoolkit/flt/FactSheets/Infrastructure/REVERSIBLE%20LANES.htm

Chang, J., Chang, K., & Chen, D. (2006). Application of 3D Laser Scanning on Measuring Pavement Roughness. *J. of Testing and Evaluation ASTM, 34*(2).

Charbonnier, P., & Nicolle, P. (2002). Semi-automatic measurement of pavement width. *Bulletin des Laboratories des Ponts et Chaussees, 235*, 85 - 94.

Chen, Liang-Chien, Shao, Yi-Chen, Jan, Huang-Hsiang, Huang, Chen-Wei, & Tien, Yong-Ming (2006). Measuring System for Cracks in Concrete Using Multitemporal Images. *Journal of Surveying Engineering ASCE, 132*(2), 77 - 82.

Cheng, H. D., & Glazier, C. (2007) *Automated Real-Time Pavement Crack Detection and Classification System* (No. 106): Transportation Research Board.

Cobb, L. (2009). Future Vision of the DHMS - personal communication.

Cobb, Lincoln (2009). Personal Communication.

CoolSuperCars (2009). Opel Insignia To Recognize Traffic Signs, from http://www.coolsupercars.com/opel-insignia-to-recognize-traffic-signs/

Corrugated Steel Pipe Institute. Guiderails, from http://www.cspi.ca/english_files/_handbook/chapter12.pdf

Council, Forrest M., Harkey, David L., Carter, Daniel L., & White, Bryon (2007). *Model Minimum Inventory of Roadway Elements - MMIRE* (No. FHWA-HRT-07-046): Federal Highway Administration.

CT DOT (2003a). Glossary, from http://www.ct.gov/dot/lib/dot/documents/dpublications/highway/Glossary.pdf

CT DOT (2003b). Roadside Safety, from http://www.ct.gov/dot/lib/dot/documents/dpublications/highway/Chapter_13.pdf

Day, Dwight (2008). *Analysis of Road Condition Data Employing Sensor Fusion and High Level Classification* (No. Proposal): Kansas State University.

Debaillon, Chris, Carlson, Paul J., He, Yefei, Schnell, Tom, & Aktan, Fuat (2007). *Updates to Research on Recommended Minimum Levels for Pavement Marking Retroreflectivity to Meet Driver Night Visibility Needs* (No. FHWA-HRT-07-059).

DOT (2000). *Primer: GASB 34.*

Dynatest. Digital Higway Data Vehicle Retrieved June 25, 2009, from http://www.pavements.com/downloads/2007aug/Digital%20Highway_v2%20small.pdf

Dynatest (2009). Road Surface Profilometer®, from http://www.pavements.com/hardware/rsp.htm

Engineering Policy Guide (April 3, 2009). Bridge Approach Slabs, from http://epg.modot.org/index.php?title=Category:503_Bridge_Approach_Slabs

Ergun, Murat, Iyinam, Sukriye, & Iyinam, A. Faik (2005). Prediction of Road Surface Friction Coefficient Using Only Macro- and Microtexture Measurements. *ASCE J. of Transportation Engineering, 131*(4), 311 - 319.

Eriksson, Jakob, Girod, Lewis, Hull, Bret, Newton, Ryan, Madden, Samuel, & Balakrishnan, Hari (2008). *The Pothole Patrol: Using a Mobile Sensor Network for Road Surface Monitoring*. Paper presented at the Proceeding of the 6th international conference on Mobile systems, applications, and services. from http://delivery.acm.org/10.1145/1380000/1378605/p29-eriksson.pdf?key1=1378605&key2=4519063421&coll=GUIDE&dl=GUIDE&CFID=38066761&CFTOKEN=64564394

eRoadInfo (2009). Asset Inventory and Pavement Management Solutions, from http://www.enterroadinfo.com/rwimaging.aspx

Evans, T. (2004). *Semi-Automated Detection of Defects in Road Surfaces.*

Facet Technology Corp. *The Automated Determination of Sign and Pavement Marking Retroreflectivity from a Moving Platform.*

Facet Technology Corporation (2005). Transportation & Mapping Solutions, Asset Management, from http://www.facet-tech.com/transportation/asset_management.htm

Fang, C. Y., Fuh, C. S., Chen, S. W., & Yen, P. S. (2003). *A Road Sign Recognition System Based on Dynamic Visual Model*. Paper presented at the Proceedings of the 2003 IEEE Computer Society Conference on Computer Vision and Pattern Recognition.

FARO news (2008). Helical Scanning - perfect road mapping of Italian streets with Siteco Informatica, from http://www.3dsnorden.se/Resources/FaroNews_108.pdf

Federal Railroad Administration (2007). *U. S. DOT National Highway-Rail Crossing Inventory, Policy Procedures and Instructions for States and Railroads.*

Ferguson, R., Pratt, D., & Macintyre, I. (1999). Automated detection and classification of cracking in road pavements (RoadcrackTM). *Road & Transport Research*

Ferguson, R., Pratt, D., Tuttle, P., Macintyre, I., Moore, D., Kearney, P., Best, M., Gardner, J., Berman, M., Buckley, M., Breen, J. & Jones, R. (2003). Patent No. 6615648. USPTO.

Ferne, Brian (2008). Research experience at the Transport Research Laboratory, UK relevant to SHRP2 Project R06 Retrieved June 25, 2009, from http://onlinepubs.trb.org/onlinepubs/shrp2/FerneNDT.pdf

FHWA (June 22, 2009). Bridge Rail Guide 2005, from http://www.fhwa.dot.gov/bridge/bridgerail/index.cfm

FHWA. Chapter 5. Driveway Crossings, from http://www.fhwa.dot.gov/environment/sidewalk2/sidewalks205.htm

FHWA (03/22/07). Clear Zone and Horizontal Clearance - Frequently Asked Questions, from http://www.fhwa.dot.gov/programadmin/clearzone.cfm

FHWA (2/6/09). FHWA Corporate Research and Technology: Rumble Strips Retrieved June 24, 2009, from http://www.fhwa.dot.gov/crt/lifecycle/rumblestrips.cfm

FHWA. Rumble Strip Types, from http://safety.fhwa.dot.gov/roadway_dept/pavement/rumble_strips/rumble_types/index.cfm

FHWA (1990). *Concrete Pavement Joints* (No. Technical Advisory T 5040.3): Federal Highway Administration.

FHWA (1995). *Recording and Coding Guide for the Structure Inventory and Appraisal of the Nation's Bridges* (No. FHWA-PD-96-001). Washington D.C.

FHWA (2001). *Roadway Shoulder Rumble Strips* (No. Technical Advisory T5040.35).

FHWA (2003a). Distress Identification Manual for the Long-Term Pavement Performance Program Available from http://www.tfhrc.gov/pavement/ltpp/reports/03031/index.htm

FHWA (2003b). *Manual on Uniform Traffic Control Devices.*

FHWA (2004). *Traffic Data Quality Measurement, Report prepared by Battelle for FHWA* from http://www.itsdocs.fhwa.dot.gov/JPODOCS/REPTS_TE/14058_files/chap4.htm.

FHWA (2006). *High Performance Concrete Pavements Project Summary: Chapter 26. NEW HAMPSHIRE HPCP PROJECT*: FHWA.

FHWA (2007a). *Model Minimum Inventory of Roadway Elements—MMIRE*: Federal Highways Administration, U.S. Department of Transportation.

FHWA (2007b). New MUTCD Sign Retroreflectivity Requirements, FHWA-SA-07-020, from http://safety.fhwa.dot.gov/roadway_dept/night_visib/policy_guide/fhwasa07020/fhwasa07020.pdf

FHWA (2007c). *Railroad-Highway Grade Crossing Handbook - Revised Second Edition.*

FHWA (2008a). HPMS Field Manual, Appendix E: Measuring Pavement Roughness Available from http://www.fhwa.dot.gov/ohim/hpmsmanl/appe.cfm

FHWA (2008b). *North American Joint Positive Train Control System Four-Quadrant Gate Reliability Assessment* (No. RR08-01).

FHWA (2009a). Digital Highway Measurement System, from http://www.volpe.dot.gov/sbir/sol05/docs/trb05handout.pdf

FHWA (2009b). MMIRE Webconference - April Retrieved June 25, 2009, from http://www.mmire.org/collateral/04-2009_webinar/MMIRE_Webconference_4_Elements.ppt#360,28,Slide 28

FHWA (2009c). TFHRC - Intersection Safety Retrieved June 30, 2009, from http://www.tfhrc.gov/safety/intersect.htm

FHWA (2009d). TFHRC - Safety Research Retrieved June 30, 2009, from http://www.tfhrc.gov/safety/index.htm#Intersections

Fletcher, Luke, Petersson, Lars, Barnes, Nick, Austin, David J., & Zelinsky, Alexander (2005). *A Sign Reading Driver Assistance System Using Eye Gaze.* Paper presented at the Proceedings of the 2005 IEEE International Conference on Robotics and Automation, Barcelona, Spain.

Fleyeh, H., Gilani, S.O., & Dougherty, M. (2006). *Road Sign Detection and Recognition using Fuzzy ARTMAP: A Case Study Swedish Speed-Limit Signs.* Paper presented at the Artificial Intelligence and Soft Computing. from http://www.actapress.com/PaperInfo.aspx?PaperID=28187&reason=500

Fleyeh, Hasan (2008). *Traffic and Road Sign Recognition.* Unpublished PhD, Napaier University.

Florida DOT (2002). How Florida DOT performs condition survey of its pavements, from http://nersp.nerdc.ufl.edu/~tia/5837-3b.pdf

Forest, R., & Utsi, V. (2004). *Non destructive crack depth measurements with Ground Penetrating Radar.* Paper presented at the IEEE Tenth International Conference on Ground Penetrating Radar.

Forkenbrock, David J., & March, Jim (2005). Issues in the Financing of Truck-Only Lanes. *Public Roads, 69*(2). Retrieved from http://www.tfhrc.gov/pubrds/05sep/02.htm

Franklin Regional Council of Governments. Roadside Barriers, from http://www.frcog.org/pubs/transportation/DesignAlternatives/ch7.PDF

Friction, Norsemeter (2009). Basic Product Features, from http://www.norsemeter.no/Products/

Geo-3D (2009). Trident-3D Road Mobile Mapping Solution, from http://www.geo-3d.com/solution.html

GeoAutomation (2009). http://www.geoautomation.be/media/geoautomation_en.pdf.

GIE (2009). GIE Products, from http://www.gieinc.ca/main_en.html

Gilani, Syed Hassan (2007). *Road Sign Recognition based on Invariant Features using Support Vector Machine.* Unpublished Master Thesis, University of Dalama.

Governments, Franklin Regional Council of. Chapter 7. Roadside Barriers, from http://www.frcog.org/pubs/transportation/DesignAlternatives/ch7.PDF

Halcrow-Group-Limited (2007). *SCANNER Surveys for Local Roads - Technical Requirements for SCANNER Survey Data and Quality Assurance (DRAFT v08)*.

Hallmark, S. L., Mantravadi, K., Veneziano, D., & Souleyrette, R. R. (2001). *Evaluating Remotely Sensed Images for use in Inventorying Roadway Infrastructure*: Center for Transportation Research and Education, Iowa State University.

Hallmark, S., Mantravadi, K., Souleyrette, R.R., & Veneziano, D. (2001). *Use of remote sensing for collection of data elements for linear refrerencing systems*: Iowa Department of Transportation.

Harwood, D. W., Council, F. M., Hauer, E., Hughes, W. E., and Vogt, A. (2000). *Prediction of the Expected Safety Performance of Rural Two-Lane Highways* (No. FHWA-RD-99-207).

Harwood, D. W., K. M. Bauer, I. B. Potts, D. J. Torbic, K. R. Richard, E. R. Kohlman Rabbani et al. (2002). *Safety Effectiveness of Intersection Left- and Right-Turn Lanes* (No. FHWA-RD-02-089): Federal Highway Administration.

Hatzidimos, John (2004). *Automatic Traffic Sign Recognition in Digital Images*. Paper presented at the Proceedings of the International Conference on Theory and Applications of Mathematics and Informatics - ICTAMI 2004, Thessaloniki, Greece.

Herold, M., & Roberts, D. (2005). Spectral characteristics of asphalt road aging and deterioration: implications for remote-sensing applications. *Applied Optics, 44*(20).

Heron, M. P., Hoyert, D. L., Murphy, B. S., Xu, J., Kochaneck, M. A., & Tejada-Vera, B. (2009). *Deaths: Final data for 2006* (No. DHHS Publication No. (PHS) 2009-1120): Centers for Disease Control and Prevention, U.S. Department of Health and Human Services.

Holick, Andrew J., & Carlson, Paul J. (2008). *Minimum Retroreflectivity Levels for Blue and Brown Traffic Signs*: Federal Highway Administration.

Hu, Fengxuan (2006). *Development and Evaluation of an Inertial Based Pavement Roughness Measuring System*. Unpublished PhD dissertation, University of South Florida.

Hu, Z., & Tsai, Y. (2009). Homography-Based Vision Algorithm for Traffic Sign Attribute Computation. *Computer-Aided Civil and Infrastructure Engineering, 24*, 1-16.

Huang, Y., & Xu, B. (2006). Automatic inspection of pavement cracking distress. *Journal of Electronic Imaging 15*(1).

Hunter, W. W., Stewart, J. Richard, Stutts, Jane C., Huang, Chen-Wei, & Pein, Wayne E. (1999). *Bicycle Lanes versus Wide Curb Lanes: Operational and Safety Findings and Countermeasure Recommendations* (No. FHWA-RD-99-035): Federal Highway Administration.

IA DOT (1999). Medians, from http://www.iowadot.gov/design/dmanual/03e-01.pdf

IBEO (2008). ALASCA, from http://www.ibeo-as.com/

IL DOT (2006). Bureau of Local Roads & Streets: Sight Distance Retrieved June 24, 2009, from http://www.dot.state.il.us/bl-/manuals/Chapter%2028.pdf

Infrastructure Management Services (2009). Data Collection Services, from http://www.ims-rst.com/data-collection.shtml

INO (2009a). http://www.ino.ca/medias/pdfs/publications/technical/3D-sensors/LCMS_Laser_Crack_Measurement_System_PRM-080043.pdf

INO (2009b). http://www.ino.ca/medias/pdfs/publications/technical/3D-sensors/LRIS_Laser_Road_Imaging_System_PRM-080048.pdf

INO (2009c). http://www.ino.ca/medias/pdfs/publications/technical/3D-sensors/LRMS_Laser_Rut_Measurement_System_PRM-080049.pdf

Integrated Publishing. Chapter 11 - Horizontal and Vertical Cuves Retrieved June 23, 2009, from http://www.tpub.com/inteng/11_htm

Integrated Publishing. Elements of a Horizontal Curve Retrieved June 24, 2009, from http://www.tpub.com/inteng/11a_htm

Interactive, Pavement (2009, May 13, 2009). Asphalt, from http://pavementinteractive.org/index.php?title=Asphalt

International Cybernetics (2009). ICC Products, from http://www.internationalcybernetics.com/products_htm

Ishak, Khairul Anuar, Sani, Maizura Mohd, Tahir, Nooritawati Md, Samad, Salina Abdul, & Hussain, Aini (2006). *A Speed limit Sign Recognition System Using Artificial Neural Network*. Paper presented at the 4th Student Conference on Research and Development (SCOReD 2006). from http://ieeexplore.ieee.org/stamp/stamp.jsp?arnumber=04339324

Ishikawa, Kiichiro, Takiguchi, Jun-ichi, Amano, Yoshiharu, & Hashizume, Takumi (2006). *A Mobile Mapping System for road data capture based on 3D road model*. Paper presented at the Proceedings of the 2006 IEEE International Conference on Control Applications, Munich, Germany.

Iteris (2009). Innovation for Better Mobility, from www.iteris.com

ITRE. Excerpts from the 2006 NCDOT Maintenance Condition Assessment Manual, from http://www.itre.ncsu.edu/NCassetMgmtConf/downloads/ExpoSupplementalInfoRoadside041808.doc

Janisch, David (2003). *An Overview of Mn/DOT's Pavement Condition Rating Procedures and Indices*: Mn DOT.

Jauregui, David, Tian, Yuan, & Jiang, Ruinian (2006). *Photogrammetry Applications in Routine Bridge Inspection and Historic Bridge Documentation* (No. NM04STR-01): New Mexico State University.

Javidi, B., Stephens, J., Kishk, S., Naughton, T., McDonald, J., & Isaac, A. (2003). *Pilot for Automated Detection and Classification of Road Surface Degradation Features* (No. JHR 03- 293): Univ. of Connecticut.

Jengo, C., Hughes, D, Veigne, J. D., & Curtis, I. (2005). *Pothole Detection and Road Condition Assessment Using Hypersectral Imagery*. Paper presented at the ASPRS 2005 Annual Conference.

Kaempchen, N., & Dietmayer, K. (2004). *Fusion of Laser Scanner and Video for Advanced Driver Assistance Systems*. Paper presented at the IEEE Conference on Intelligent Transportation Systems. from http://www.uni-ulm.de/fileadmin/website_uni_ulm/iui.inst.110/Downloads/Publikationen/2004/kaempchen04e.pdf

Karamihas, Steven M. (2005). *Critical Profiler Accuracy Requirements*: University of Michigan Transportation Research Institute Report (UMTRI-2005-24).

Kastrinaki, V., Zervakis, M., & Kalaitzakis, K. (2003). A survey of video processing techniques for traffic applications. *Image and Vision Computing, 21*, 359-381.

Kayama, K., Yairi, I. E., & Igi, S. (2007). *Detection of Sidewalk Border using Camera on Low-Speed Buggy*. Paper presented at the Proceedings of Artificial Intelligence and Applications, Innsbruck, Austria.

Kelly-Creswell Company (2009). Raised Pavement Marker Detector, from
http://www.kellycreswell.com/pages/electronics/RPMD.htm

Kentucky Lifesavers Conference (2005). Geometric Features Related to Substantive Safety at Intersections, June 24, 2009, from http://lifesavers.ky.gov/lifesavers_2006/session22-ranck2.ppt#510,1,Geometric Features Related to Substantive Safety at Intersections

Kim, Seung-Hun, Roh, Chi-Won, Kang, Sung-Chul, & Park, Min-Yong (2007). *Outdoor Navigation of a Mobile Robot Using Differential GPS and Curb Detection*. Paper presented at the 2007 IEEE International Conference on Robotics and Automation. from http://imm.kist.re.kr/Paper/Conf/2007-Outdoor%20Navigation%20of%20a%20Mobile%20Robot%20Using%20Differential%20GPS%20and%20Curb%20Detection.pdf

Kodagoda, K. R. S., Ge, S. S., Wijesoma, W. S., & Balasuriya, A. P. (2007). IMMPDAF Approach for Road-Boundary Tracking. *IEEE Transactions on Vehicular Technology, 56*(2), 478 - 496.

Kodagoda, K.R.S., Wang, C-C., & Dissanayake, G. (2005). *LASER-BASED SENSING ON ROADS*. Paper presented at the Proceedings of the Intelligent Vehicles and Road Infrastructure Conference, Melbourne, Australia.

Koutaki, G., & Uchimura, K. (2003) *Automatic road extraction based on cross detection in suburb*. Paper presented at the SPIE. from http://navi.cs.kumamoto-u.ac.jp/english/publications/pdf/int_thesis/2004-02.pdf

Kremer, J., and Hunter, G. *Performance of the StreetMapper Mobile LIDAR Mapping System in "Real World" Projects*.

Lane county Public Works. Appendix D: Detailed Level of Service Methodology, from
http://www.lanecounty.org/Departments/PW/TransPlanning/Pages/default.aspx

Laurent, J., Lefebvre, D., & Samson, E. (2008). *Development of a New 3D Transverse Laser Profiling System for the Automatic Measurement of Road Cracks*. Paper presented at the Proceedings of the 6th Symposium on Pavement Surface Characteristics – SURF 2008. from http://www.ino.ca/medias/pdfs/publications/scientific/LCMS_Surf_2008.pdf?URI=oe-16-7-4638

Laurent, John, & Hebert, Jean-Francois (2002). *High Performance 3D Sensors for the characterization of Road Surface Defects*. Paper presented at the IAPR Workshop on Machine Vision Applications. from http://www.cvl.iis.u-tokyo.ac.jp/mva/proceedings/CommemorativeDVD/2002/papers/2002388.pdf

Lay, M. G. (1990). *Handbook of Road Technology* (Second ed. Vol. Volume 2 Traffic and Transport): Gordon and Breach Science Publishers.

Lee, Byoung Jik, & Lee, Hosin "David" (2004). Position-Invariant Neural Network for Digital Pavement Crack Analysis. *Computer-Aided Civil and Infrastructure Engineering, 19*, 105 - 118.

Lee, Hoon, Park, Soonyoung, & Choi, Kyoungho (2009). Support Vector Machines For Understanding Lane Color and Sidewalks. *PROCEEDINGS OF WORLD ACADEMY OF SCIENCE, ENGINEERING AND TECHNOLOGY, 38*, 1053 - 1056.

Leng, S-S., Vrignon, J., Gruyer, D., & Aubert, D. (2005). *A New Multi-Lanes Detection Using Multi-Camera for Robust Vehicle Location*. Paper presented at the Proc. Of IEEE Intelligent Vehicles Symposium.

Levinson, David. Vertical Curves Retrieved June 24, 2009, from http://nexus.umn.edu/Courses/ce3201/CE3201-L3-06.pdf

Li, Qing, Zheng, Nanning, & Cheng, Hong (2004). Springrobot: A Prototype Autonomous Vehicle and Its Algorithms for Lane Detection. *IEEE Transactions on Intelligent Transportation Systems, 5*(4), 300 - 308.

Liou, Yi-Sheng, Duh, Der-Jyh, Chen, Shu-Yuan, & Hsieh, Jun-Wei (2005). A Fast Method to Detect and Recognize Scaled and Skewed Road Signs *Advanced Concepts for Intelligent Vision Systems* (pp. 68 - 75). Berlin / Heidelberg: Springer

Liu, Wei (2009). Automatic Road Sign Recognition From Video http://www.slideshare.net/paveman/Automatic-Road-Sign-Recognition-from-Video.

Lu, J. (2002). *Development of an Automatic Detection System for Measuring Pavement Crack Depth on Florida Roadways* (No. BB-884).

Lu, Xiaoye, & Manduchi, Roberto (2005). *Detection and Localization of Curbs and Stairways Using Stereo Vision.* Paper presented at the 2005 IEEE International Conference on Robotics and Automation, Barcelona, Spain.

Maerz, Norbert H., & Niu, Qiang (2003). *Automated Mobile Highway Sign Visibility Measurement System.* Paper presented at the Transportation Research Board, 82th Annual Meeting from http://web.mst.edu/~norbert/pdf/trb02_maerz_sign.pdf

Maldonado-Bascón, Saturnino, Lafuente-Arroyo, Sergio, Gil-Jiménez, Pedro, Gómez-Moreno, Hilario, & López-Ferreras, Francisco (2007). Road-Sign Detection and Recognition Based on Support Vector Machines. *IEEE Transactions on Intelligent Transportation Systems, 8*(2), 264 - 278.

Mandli (2009). Systems - technology for the road and rail, from http://www.mandli.com/systems/systems.php

Mason Academic Research System. Vertical Curves and Horizontal Sight Distance Retrieved June 24, 2009, from http://mason.gmu.edu/~aflanner/CEIE_360/CHAPTER%203%20part%20II%20st.ppt#292,18,Stopping Sight Distance & Crest Curves

McCall, Joel C., & Trivedi, Mohan M. (2006). Video-Based Lane Estimation and Tracking for Driver Assistance: Survey, System, and Evaluation. *IEEE Transactions on Intelligent Transportation Systems, 7,* 20 - 37.

McGhee, Kevin K. (2000). *Quality Assurance of Road Roughness Measurement*: Virginia Transportation Research Council (VTRC 00-R20). Virginia Department of Transportation.

Messina, E., Jacoff, A., Scholtz, J., Schlenoff, C., Huang, H., Lytle, A., Blitch, J. (2005). *Statement Of Requirements For Urban Search and Rescue Robot Performance Standards*. Gaithersburg, MD: National Institue of Standards and Technology.

MN DOT (2006). *Roadway Lighting Design Manual*.

Mobileye (2009). EyeQ™, Mobileye's System-on-Chip, from http://www.mobileye.com/

Moutarde, Fabien, Bargeton, Alexandre, Herbin, Anne, & Chanussot, Lowik (2007). *Robust on-vehicle real-time visual detection of American and European speed limit signs, with a modular Traffic Signs Recognition system.* Paper presented at the Proceedings of the 2007 IEEE Intelligent Vehicles Symposium, Istanbul, Turkey.

Mraz, Alexander (2004). *Evaluation of Digital Imaging Systems Used in Highway Applications.* Unpublished PhD Thesis, University of South Florida.

Muller-Schneiders, Stefan, Nunn, Christian, & Meuter, Mirko (2008). *Performance Evaluation of a Real Time Traffic Sign Recognition System.* Paper presented at the IEEE Intelligent Vehicles Symposium.

Mullis, C., Reid, J., Brooks, E., & Shippen, N. (2005). *Automated Data Collection Equipment for Monitoring Highway Condition* (No. SPR 332 and FHWA-OR-RD-05-10): Oregon DOT and FHWA.

Mustaffar, M., Ling, T. C., & Puan, O.C. (2008). Automate Pavement Imagin Program APIP) for Pavement Cracks Classification and Quantification - A Photogrammetric Approach. *The International Archives of the Photogrammetry, Remote Sensing and Spatial Information Sciences, XXXVII*(B4).

Nakamura, Mitsuaki (2001). US Patent 6185338 Patent No.

NCDOT. Inventory Asset Data Collection Elements. *National Conference: Highway Asset Inventory and Data Collection*, from http://www.itre.ncsu.edu/ncassetmgmtconf/downloads/RoadsideDataElementTable.pdf

NCDOT. NCDOT Pavement Condition Survey. *National Conference: Highway Asset Inventory & Data Collection*, 2009, from http://www.itre.ncsu.edu/ncassetmgmtconf/downloads/PavementsDataElementTable.pdf

Nedevschi, Sergiu, Schmidt, Rolf., Graf, Thorsten, Danescu, Radu, Frentiu, Dan, Marita, Tiberiu, et al (2004). *3D Lane Detection System Based on Stereovision*. Paper presented at the 2004 IEEE Intelligent Transportation Systems Conference. from http://cv.utcluj.ro/Publications/ITSC2004.pdf

Nehate, Girish, & Rys, Malgorzata (2006). 3D Calculation of Stopping-Sight Distance from GPS Data *Journal of Transportation Engineering ASCE, 132*(9), 691 - 698.

NHTSA (2000). *Traffic Safety Facts 2000*. Washington, DC: National Highway Traffic Safety Administration, U.S. Department of Transportation.

NHTSA (2009). *Traffic Safety Facts 2007*. Washington, DC: National Highway Traffic Safety Administration, U.S. Department of Transportation.

North Carolina Department of Transportation (2008). 2008 Highway Asset Inventory & Data Collection, Roadside Manual and Vendor Data Sets Retrieved June 25, 2009, from http://www.itre.ncsu.edu/NCassetMgmtConf/vendordatasets.html

Noyce, D. A., Bahia, H. U., Yambó. J. M., & Kim, G. (2005). Incorporating road safety into pavement management: Maximizing asphalt pavement surface friction for road safety improvements. *Midwest Regional University Transportation Center Traffic Operations and Safety (TOPS) Laboratory, Draft Literature Review and State Surveys (April 2005)*.

Offrell, P., Sjogren, L., & Magnussen, R. (2005). Repeatability in Crack Data Collection on Flexible: Comparison between Surveys Using Video Cameras, Laser Cameras, and a Simplified Manual Survey. *ASCE J. of Transportation, 131*(1), 552-562.

Oniga, Florin, Nedevschi, Sergiu, & Meinecke, Marc Michael (2007). *Curb Detection Based on Elevation Maps from Dense Stereo*. Paper presented at the IEEE International Conference on Intelligent Computer Communication and Processing. from http://ieeexplore.ieee.org/xpls/abs_all.jsp?arnumber=4352150

Optech (2009). Optech's LYNX Mobile Mapper, from http://www.optech.ca/pdf/LynxDataSheet.pdf

Otsuka, Y., Muramatsu, S., Takenaga, H., Kobayashi, Y., & Monji, T. (2002). Multitype Lane Markers Recognition Using Local Edge Direction. *Proceedings of the IEEE Intelligent Vehicles Symposium, 2*, 604 - 609.

Otsuka, Yuji, Muramatsu, Shoji, Takenaga, Hiroshi, Takezaki, Jiro, & Monjii, Tatsuhiko (2007). US Patent No. 7295682.

Paclík, Pavel, Novovicova, Jana, & Duin, Robert P. W. (2006). Building Road-Sign Classifiers Using a Trainable Similarity Measure. *IEEE Transactions on Intelligent Transportation Systems, 7*(3).

Pathway Services Inc. (2009). Data Collection: Services and Vehicles, from http://www.pathwayservices.com/index.html

Pidwerbesky, B., Waters, J., Gransberg, D., & Stemprok, R. (2006). *Road surface texture measurement using digital image processing and information theory* (No. Report 290): Land Transport New Zealand Research.

Plaxico, C. A., Ray, M. H., Weir, J. A., Orengo, F., Tiso, P., McGee, H., et al. (2005). *Recommended Guidelines for Curb and Curb-Barrier Installations* (No. NCHRP Report 537).

PrecisionScan (2009). Mobile surveys of roadway visibility and color, from http://www.precisionscan.com/

Rajab, Maher, Alawi, Mohammad, & Saif, Mohammed (2008). Application of Image Processing to Measure Road Distress. *WSEAS Transactions on Information Science and Applications, 5*(1), 1-7.

Ramboll (2009). Services, from http://intl.rambollrst.se/services/index.shtml#s8

RoadVista (2009). Mobile Instruments: Laserlux® CEN 30 for Pavement Markings, from http://www.roadvista.com/products/laserluxcen30.shtml

Roadware (2009). Fugro Roadware - ARAN, from http://www.roadware.com/products_services/aran/

Roberts, D., Gardner, M., Funk, C., & Nornha, V. (2001). *Road Extraction Using Spectral Mixture and Q-Tree Filter Techniques*: University of California, Santa Barbara.

Rodegerdts, Lee A., Nevers, Brandon, Robinson, Bruce, Ringert, John, Koonce, Peter, Bansen, Justin, et al. (2004). *Signalized Intersections: Informational Guide* (No. FHWA-HRT-04-091).

Ross, H. E. Jr., Sicking, D. L., Zimmer, R. A., & Michie, J. D. (1993). *NCHRP Report 350: Recommended Procedures for the Safety Performance Evaluation of Highway Features.*

Rumar, K (1985). The role of perceptual and cognitive filters in observed behavior. *Human behavior and traffic safety*, 151-165.

Schlenoff, Craig, Messina, Elena, Lytle, Alan, Weiss, Brian, & Virts, Ann (2007). Test Methods and Knowledge Representation for Urban Search and Rescue Robots In Houxiang Zhang (Ed.), *Climbing & Walking Robots, Towards New Applications*: I-Tech Education and Publishing.

Schlosser, Jeffrey (2007). Efficient Traffic Sign Detection Using 3D Scene Geometry. Standford University.

Schniering Ing (2009). Pavement Monitoring by the ARGUS Survey System, from http://www.schniering.com/Englisch/index_eng.php

Schofield, K., & Lynham, N. (2009). 7526103. USPTO.

SDDOT (2007). Transportation Inventory Management / Pavement Condition Monitoring. Retrieved 05/08/2009, from South Dakota Department of Transportation: http://www.sddot.com/pe/data/pave.asp

Seng, John (2008). *Sidewalk Following Using Color Histograms*. Paper presented at the The 2008 Consortium for Computer Sciences in the Colleges - Southwestern Regional Conference. from http://users.csc.calpoly.edu/~jseng/ccsc_paper.pdf

Shafiq, Tahir. Productivity Gains in Mobile Mapping: Applanix Landmark Retrieved June 25, 2009, from http://www.gisdevelopment.net/technology/gps/ma07143.htm

Shirvaikar, Mukul V. (2004). *Automatic Detection and Interpretation of Road Signs*. Paper presented at the Proceedings of the Thirty-Sixth Southeastern Symposium on System Theory.

Shiyab, Adnan (2007). *Optimum use of the flexible pavement condition indicators in pavement management system.* Curtin University of Technology.

Shneier, Michael (2005). *Road Sign Detection and Recognition*. Paper presented at the IEEE Computer Society International Conference on Computer Vision and Pattern Recognition. from http://www.isd.mel.nist.gov/documents/shneier/Road_Sign_Detection.pdf

Siegmann, Philip, López-Sastre, Roberto Javier, Gil-Jiménez, Pedro, Lafuente-Arroyo, Sergio, & Maldonado-Bascón, Saturnino (2008). Fundaments in Luminance and Retroreflectivity Measurements of Vertical Traffic Signs Using a Color Digital Camera. *IEEE Transaction on Instrumentation and Measurement, 57*(3), 607 - 615.

Silapachote, Piyanuch, Weinman, Jerod, Hanson, Allen, Weis, Richard, & Mattar, Marwan A. (2005). *Automatic Sign Detection and Recognition in Natural Scenes*. Paper presented at the Proceedings of the 2005 IEEE Computer Society Conference on Computer Vision and Pattern Recognition (CVPR'05). from ftp://vis-ftp.cs.umass.edu/Papers/silapachote/silapachote05automatic.pdf

Sokolic, Ivan (2003). *Criteria to Evaluate the Quality of Pavement Camera Systems in Automated Evaluation Vehicles*. Unpublished Masters Thesis.

Stein, William J., & Neuman, Timothy R. (2007). *Mitigation Strategies for Design Exceptions* (No. FHWA-SA-07-011).

Sun, Wei, Chen, Zuemin, Chen, Yuanhang, Ekbote, Aditya, & Liu, Richard Ce (2006). *Auto-synchronized laser scanning range sensor for thermoplastic pavement marking material thickness measurement*. Paper presented at the Proceedings of the SPIE Conference on Smart Structures and Materials. from http://cat.inist.fr/?aModele=afficheN&cpsidt=18359939

Surface Systems & Instruments (2009). Pavement Management Solutions, from http://www.smoothroad.com/

Swargam, Nagajyothi (2004). *Development of a Neural Network Approach for the Assessment of the Performance of Traffic Sign Retroreflectivity* Unpublished Master Thesis, Louisiana State University.

Technology Enhanced Learning. Section 2: Linear Referencing Systems Retrieved June 24, 2009, from http://ecow.engr.wisc.edu/cgi-bin/get/cee/659/adams/readings/linearloca/section2.pdf

Teomete, Egemen, Amin, Viren R., Ceylan, Halil, & Smadi, Omar (2005). *Digital Image Processing for Pavement Distress Analyses*. Paper presented at the Proceedings of the 2005 Mid-Continent Transportation Research Symposium, Ames, Iowa.

Terrametrix (2009). Asset Inventory. from http://www.terrametrix3d.com/

Texas A&M University. Traffic Signals Retrieved June 24, 2009, from http://fcs.tamu.edu/safety/passenger_safety/toolkit/roadway_markings/traffic_signals_read_your_road.pdf

Thomas, Gary B., & Schloz, Courtney (2001). *Durable, Cost-Effective Pavement Markings Phase I: Synthesis of Current Research*: Center for Transportation Research and Education, Iowa State University.

Timm, David H., & McQueen, Jason M. (2004). *A study of manual vs. automated pavement condition surveys* (No. Report IR-04-01): Auburn University.

Topcon (2009). Integrated Positioning System for 3D Mobile Mapping, from http://www.topconpositioning.com/products/mapping-and-gis/mobile-mapping/ip-s2.html

Torresen, J., Bakke, J. W., & Sekanina, L. (2004). *Efficient Recognition of Speed Limit Signs*. Paper presented at the IEEE Intelligent Transportation Systems Conference.

Transport Canada (2008, 2008-12-01). Structural Defects Found in Asphaltic Concrete Pavements from http://www.tc.gc.ca/civilaviation/international/Technical/Pavement/quality/structural/asphalticConcrete.htm#map

Transport Research Laboratory (2009). Highways Agency Road Research Information System (HARRIS), from http://www.trl.co.uk/facilities/mobile_test_equipment/highways_agency_road_research_information_system.htm

Transportation Engineering Online Lab Manual. Geometric Design Glossary Retrieved June 24, 2009, from http://www.webs1.uidaho.edu/niatt_labmanual/Chapters/geometricdesign/Glossary/index.htm

TRB (2004). *NCHRP Synthesis 334 Automated Pavement Distress Collection Techniques*.

TRB (2008). *SHRP-2 Safety Program Brief*: Transportation Research Board, National Academy of Sciences.

TRB (2009a). *Crash standards save lives, spur innovation*: Transportation Research Board, National Academy of Sciences.

TRB (2009b). Roadway Measurement System Evaluation Retrieved April 17, 2009, from http://www.trb.org/TRBNet/ProjectDisplay.asp?ProjectID=2146

Trentacoste, M. F. (2006). Digital Highway Measurement System (Presentation to the 5th International Visualization In Transportation Symposium). McLean, VA: Office of Safety Research and Development, Turner Fairbank Highway Research Center, FHWA.

Tsai, James (2007). Enhanced Transportation Asset Data Collection Using Video Log Image Pattern Recognition Retrieved March 26, 2009, from http://www.gis-t.org/files/SenCV.pdf

Tsai, Yichang (James), Wu, Mingzhan, & Wang, Zhaohua (2004). *Re-Engineered Roadway Transportation Data Inventory Using GPS/GIS*. Paper presented at the Proceedings of 8th International Conference on Applications of Advanced Technologies in Transportation Engineering.

Tsai, Yichang James, Wu, Jianping, Wu, Yiching, & Wang, Zhaohua (2005). Automatic Roadway Geometry Measurement Algorithm Using Video Images *Image Analysis and Processing – ICIAP 2005* (pp. 669 - 678). Berlin / Heidelberg: Springer.

Tsai, Yichang, Mingzhan Wu, & Zhaohua Wang (2004, May 26-28). *Re-engineered Roadway Transportation Data Inventory Using GPS/GIS*. Paper presented at the Proceedings of the Eighth International Conference, Applications of Advanced Technologies in Transportation Engineering (2004), Beijing, China.

Turan, Jan, Turan Jr., Jan, Ovsenik, Lubos, & Fifik, Martin (2008). *Architecture of Invariant Transform Based Traffic Sign Recognition System*. Paper presented at the IEEE Proceedings of the 18th International Conference Radioelektronika.

Vandervalk, Anita (2008, July 17, 2008). *SHRP 2 Project S-03, Task 2, Determination & Prioritization of Data Elements*. Paper presented at the Fourth SHRP Safety Symposium.

Veit, Thomas, Tarel, Jean-Philippe, Nicolle, Philippe, & Charbonnier, Pierre (2008). *Evaluation of Road Marking Feature Extraction*. Paper presented at the Proceedings of the 11th International IEEE Conference on Intelligent Transportation Systems. from http://perso.lcpc.fr/tarel.jean-philippe/publis/itsc08.html

Veneziano, D., Souleyrette, R., & Hallmark, S. (2003). *Integrating LIDAR and Photogrammetry in Highway Location and Design*. Paper presented at the Transportation Research Board 2003 Annual Meeting. from http://www.ltrc.lsu.edu/TRB_82/TRB2003-000908.pdf

Veneziano, David, Hallmark, Shauna L., Souleyrette, Reginald R., & Mantravadi, Kamesh (2002). *Evaluating Remotely Sensed Images for Use in Inventorying Roadway Features*. Paper presented at the Proceedings of the Seventh International Conference of the Applications of Advanced Technologies in Transportation from http://scitation.aip.org/getabs/servlet/GetabsServlet?prog=normal&id=ASCECP000245040632000048000001&idtype=cvips&gifs=yes

Viatech (2009). http://viatech.no/ezpublish-4.0.0/index.php/eng/Products/Profile-and-texture.

Vijay, S. (2006). *Low Cost – FPGA based system for pothole detection on Indian Roads*. Indian Institute of Technology, Bombay.

Viner, H., Abbott, P., Dunford, A., Dhillon, N., Parsley, L., & Read, C. (2006). *Surface Texture Measurement on Local Roads* (No. Published Project Report 148): TRL Limited.

Wang, H. (2005). *Development of Laser System to Measure Pavement Rutting*. University of South Florida.

Wang, K. C. P., & Gong, W. (2002). *Automated Pavement Distress Survey: A Review and A New Direction*. Paper presented at the 2002 Pavement Evaluation Conference. from http://pms.nevadadot.com/2002presentations/43.pdf

Wang, Kelvin C. P., & Gong, Weiguo (2005). Real-Time Automated Survey System of Pavement Cracking in Parallel Environment. *Journal of Infrastructure Systems, ASCE*, 154 - 164.

Wang, Yue, Teoha, Eam Khwang, & Shen, Dinggang (2004). Lane detection and tracking using B-Snake. *Image and Vision Computing, 22*, 269 - 280.

Watson, P., & Wright, A. (2006). *New Pattern Recognition Methods and the Detection of Edge Deterioration* (No. Report 141): TRL Limited for Transport Research Foundation.

WayLink (2009). WayLink Products, from http://www.waylink.com/

Wijesoma, W. S., Kodagoda, K. R. S., & Balasuriya, A. P. (2005). Laser-camera Compositite Sensing for Road Detection and Tracking. *International Journal of Robotics and Automation*.

Wikipedia (May 12, 2009). Wigwag (Railroad) Retrieved June 23, 2009, from http://en.wikipedia.org/wiki/Wigwag_(railroad)

Won, Woong-Jae, Lee, Minho, & Son, Joon-Woo (2008). *Implementation of Road Traffic Signs Detection Based on Saliency Map Model*. Paper presented at the IEEE Intelligent Vehicles Symposium.

WSDOT (2009). Washington State Department of Transportation Pavement Guide, Module 9: Pavement Evaluation, Section 4: Skid Resistance Retrieved 05/08/2009, 2009, from http://training.ce.washington.edu/wsdot/Modules/09_pavement_evaluation/09-4_body.htm

Wu, J., & Tsai, Y. (2006). Enhanced Roadway Geometry Data Collection Using an Effective Video Log Image-Processing Algorithm. *Transportation Research Record: Journal of the Transportation Research Board, 1972*, 133-140.

Wu, Jianping, & Tsai, Yichang (2005). Real-time Speed Limit Sign Recognition Based on Locally Adaptive Thresholding and Depth-First-Search. *Photogrammetric Engineering and Remote Sensing, 71*(4), 405 - 414.

Wu, Wen, Chen, Xilin, & Yang, Jie (2005). Detection of Text on Road Signs From Video. *IEEE Transactions on Intelligent Transportation Systems, 6*(4), 378 - 390.

Xu, Bugao, & Huang, Yaxiong (2003). *Automated Pavement Cracking Rating System: A Summary* (No. Report 7-4975-S): University of Texas at Austin.

Xu, G., Ma, J., Liu, F., & Niu, X. (2008). *Automatic Recognition of Pavement Surface Crack Based on BP Neural Network*. Paper presented at the International Conference on Computer and Electrical Engineering.

Yen, Kin, Swanston, Travis, Hiremagalur, Jagannath, Ravani, Bahram, & Lasky, Ty (2005). *A BRIDGE-HEIGHT SENSING AND DATABASE MANAGEMENT SYSTEM FOR RELIABLE AND EFFICIENT OVERSIZE PERMITTING AND ROUTING* (No. AHMCT Research Report UCD-ARR-05-02-28-01): University of California Davis.

Yoon, Changyong, Lee, Heejin, Kim, Euntai, & Park, Mignon (2008). Real-Time Road Sign Detection Using Fuzzy-Boosting. *IEICE Transactions on Fundamentals of Electronics, Communications and Computer Sciences, E91-A*(11), 3346-3355.

Yotta (2009). Automated Surface Condition Surveys (SCANNER), from http://www.yotta.tv/surveys/scanner.php

Yu, Chunhe, & Zhang, Danping (2006). *Road Curbs Detection Based on Laser Radar*. Paper presented at the 8th International Conference on Signal Processing.

Zhang, Chunsun (2008). *An UAV-Based Photogrammetric Mapping System for Road Condition Assessment*. Paper presented at the The International Archives of the Photogrammetry, Remote Sensing and Spatial Information Sciences, Vol. XXXVII, Part 5. from http://www.isprs.org/congresses/beijing2008/proceedings/5_pdf/109.pdf

Zhou, Jian, Huang, Peisen, & Chiang, Fu-Pen (2006). Wavelet-based pavement distress detection and evaluation. *Optical Engineering, 45*(2).

Zhou, X., & Huang, X-Y. (2004). *Multi Lane Line Reconstruction for Highway Application with A Signal View*. Paper presented at the Proceedings of the Third International Conference on Image and Graphics (ICIG'04).

Zin, Thi Thi, & Hama, Hiromitsu (2004). *Robust Road Sign Recognition Using Standard Deviation*. Paper presented at the IEEE Intelligent Transportation Systems Conference.

www.ingramcontent.com/pod-product-compliance
Lightning Source LLC
Chambersburg PA
CBHW080300180526
45167CB00006B/2605